BOOK ONE

ALAN FRASER

Montrose Academy
Montrose
Angus

IAN GILCHRIST

Kirkcaldy High School
Kirkcaldy
Fife

Oxford University Press

Oxford University Press, Walton Street, Oxford OX2 6DP

Oxford London New York Toronto
Melbourne Auckland Kuala Lumpar Singapore
Hong Kong Tokyo Delhi Bombay Calcutta
Madras Karachi Nairobi Dar es Salaam
Cape Town

and associated companies in
Beirut Berlin Ibadan Mexico City Nicosia

Oxford is a trademark of Oxford University Press

© Alan Fraser, Ian Gilchrist 1985

ISBN 0 19 914235 1

Printed in Hong Kong

Contents

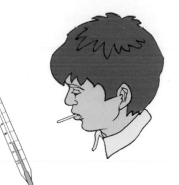

Contents

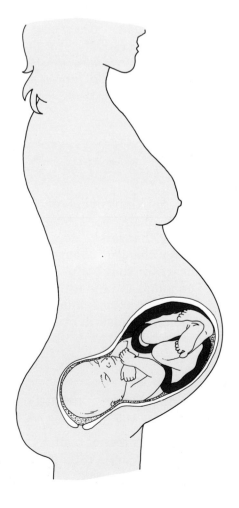

Contents

7 Electricity   *Introduction*   83

8 The gases of the air   *Introduction*   101

About this book . . . .

Your science class is a place for doing experiments. In fact, science is all about doing experiments. Day in, day out, scientists carry out experiments, trying to find the answers to many different problems. If you look at page 1, you can see some of the things they do.

But your science class is also a place for reading books. Scientists spend lots of time reading books, searching for information, looking up instructions about how to do experiments, finding out what other scientists have done. You will have to learn how to do this, too.

Starting Science has been written to be your first 'scientific book'. It has different jobs to do. It has been written to help you
- to understand what you find out in your own experiments
- to see where science fits into everyday life
- to realise how important science is, and how much scientists have been able to improve the world you live in
- to begin to think like a scientist

In writing it, we have also tried to find things which will interest you, and things which you will enjoy doing. We hope that we have succeeded!

Alan Fraser
Ian Gilchrist

And how to use it . . . .

Starting Science is made up of units. Most units contain three pages.

Starting Off is the first page. In it, you will learn a new piece of science. You *must* begin with this page. Otherwise, the other pages won't make any sense.

Going Further is the second page. It follows on from what you learned in **Starting Off**.

For The Enthusiast, the third page, takes you even further. The material on it is usually more difficult.

When you start to work on a page, you should first read everything thoroughly – including *Did You Know*. You should also look carefully at any diagrams. Then you can answer the questions. Some questions end with a triangle sign (▲). This tells you that the answer to the question is written somewhere on the page. Some questions begin **Try to find out**. You will usually have to look through other books – like encyclopaedias – for the answers to these. To answer the other questions, you will have to use what you have learned on the page, and a bit of brain power! Using your brain is all part of **Starting Science!**

Beginning to be a scientist

Lots of exciting things happen in science:

Astronomers discover new stars.

Some engineers design new aeroplanes.

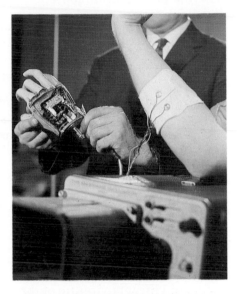

Other engineers design spare parts for the human body.

Chemists invent new drugs and medicines.

Vets and doctors use the new medicines to cure diseases.

Geologists look for – and sometimes find – precious minerals.

But on the day that you have your first science lesson, you won't do any of these things.
That shouldn't really surprise you – or disappoint you. Even the world's greatest scientists didn't make any brilliant discoveries in their first science lesson! They had to learn to be scientists, and so will you. You are going to start where all scientists have to start –

right at the beginning.

As you learn to be a scientist, you will find out more and more about the things around you.

You have five senses to help do this. They are **seeing**, **hearing**, **feeling**, **smelling** and **tasting**. But these senses may let you down. You can't always trust them to give the correct answer.

Seeing
Touching
Tasting
Smelling
Hearing

Your senses need help

You can't always depend on your eyes to judge lengths. The top line in the picture on the right looks longer than the bottom one. In fact, the lines are the same length.

Is one line longer than the other?

You can't depend on your senses to find out the temperature of the air, either. The same air may feel warm to one person and cold to another.

To find out if one line is longer than another, you can use a **ruler**. To find air temperature, you can use a **thermometer**.

The thermometer and the ruler are measuring tools called **instruments**. One of the first lessons to learn in science is this:

Don't depend on your senses only – use a measuring instrument whenever you can.

Here are some of the measurements which are made by scientists, and the measuring instruments used to make them.

LONDON AIRPORT
I'm very cold
I'm very hot

Is it warm today?

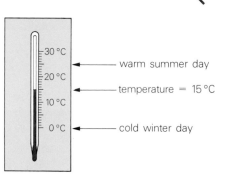

30 °C
20 °C — warm summer day
10 °C — temperature = 15 °C
0 °C — cold winter day

Scientific answer:
The temperature is 15 °C (15 degrees centigrade)

1   Time

A **stopwatch** is an instrument for measuring time.
Minutes (min) and seconds (s) are **units** of time.

You should always give a measurement as a number followed by a unit, like this:

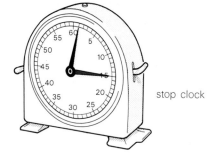
stop clock

electronic stopwatch

The girl took 1 min 15 s to run round the track
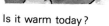
number   unit   number   unit

2 Length

Rulers, metre sticks and tape measures are used to measure length. Metres (m), centimetres (cm) and millimetres (mm) are units of length:

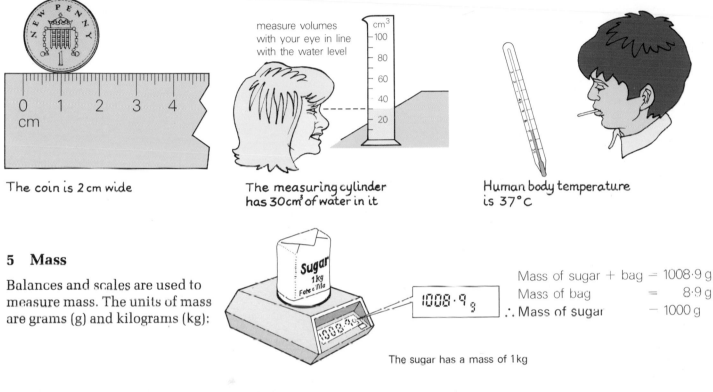

The coin is 2 cm wide

3 Volume of a liquid

Volume means 'space taken up'.

A measuring cylinder is used to measure the volume of a liquid. The cubic centimetre (cm^3) is the unit of volume used in science. Litres (l) and millilitres (ml) are other units of volume:

measure volumes with your eye in line with the water level

The measuring cylinder has $30 cm^3$ of water in it

4 Temperature

Temperature means 'hotness'.

A thermometer is used to measure temperature. The units of temperature are degrees Celsius (°C):

Human body temperature is 37°C

5 Mass

Balances and scales are used to measure mass. The units of mass are grams (g) and kilograms (kg):

Mass of sugar + bag = 1008·9 g
Mass of bag   =   8·9 g
∴ Mass of sugar   = 1000 g

The sugar has a mass of 1 kg

Did you know?

- An airline pilot cannot use his senses to guide his plane through thick cloud. He must depend on the plane's instruments.
- The highest air temperature ever recorded was 58°C, in the Libyan desert, in 1922.

1 Name your five senses. ▲
2 Why do scientists use measuring instruments whenever they can?
3 When would you use: a) a stopwatch b) a thermometer
   c) a measuring cylinder? ▲
4 Which units are used to measure: a) volume b) length
   c) mass? ▲
5 Make a list of 5 measuring instruments used in your home.
6 What are the readings on the instruments opposite?
7 **Try to find out:** something about the instruments which help pilots to fly planes through thick cloud.

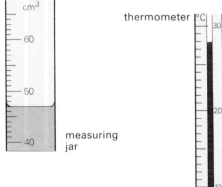

thermometer

measuring jar

stop clock

1.1 Solving measurement problems

Going further

Measuring is usually easy – if you have the correct measuring instrument. But some measurements cause problems. They need some thought.

The volume of an odd-shaped object

You can find the volume of a pebble (or any other odd-shaped object) using a measuring cylinder.

First you put some water in the measuring cylinder, and measure the volume. Then you drop in the pebble and measure the volume again. The difference in volume is equal to the volume of the object.

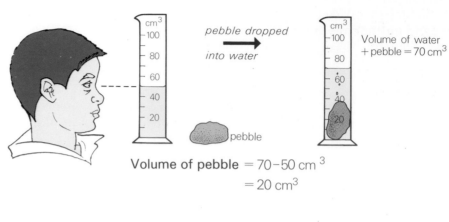

pebble dropped into water

Volume of water + pebble = 70 cm³

pebble

Volume of pebble = 70 − 50 cm³
= 20 cm³

The time of a pendulum swing

You can make a pendulum from a piece of string and an iron bolt. First, you hang the bolt from the string. Then you set the bolt swinging.

No matter how far the bolt swings, the time for one swing is always the same. One swing may be too fast to time accurately. But you can get a good answer by timing several swings.

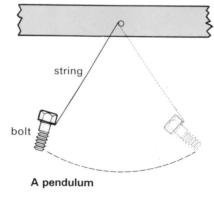

string

bolt

A pendulum

time for 10 swings = 12·0 s

each swing of the pendulum takes the same time

$\therefore$ Time for 1 swing $= \dfrac{12\,\text{s}}{10}$

$= 1\cdot2\,\text{s}$

The mass of a pin

One pin may not be enough to give a reading on the balance. But there will be a reading with 100 pins on the balance pan.

mass of 100 pins = 10·0 g

if all the pins are the same:

mass of 1 pin $= \dfrac{10\cdot0\,\text{g}}{100} = 0\cdot1\,\text{g}$

if the pins are not exactly the same
0·1 g is the *average* mass of 1 pin

1 a) What is special about a pendulum swing? ▲
   b) How can you time a pendulum swing accurately? ▲
2 How would you find the mass of a match?
3 Find the average mass of a nail if 50 nails have a mass of 175 g.
4 How would you find the volume of: a) a golf ball   b) a cork?
5 How long does it take for a midge to beat its wings once?
6 You could find your own volume with a bath of water, a measuring jug, a marker pen and a bit of help. What would you do?

Did you know?

Scientists have found out that:
- a midge insect beats its wings 1000 times each second
- a flu virus is only 0.000 001 5 cm long.

1.1 Setting our standards

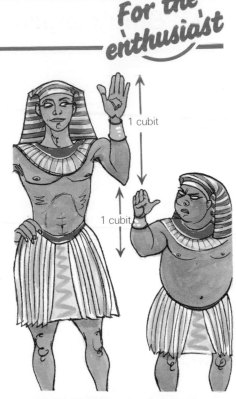

Cubits had different lengths !

In the days before measuring instruments were invented, people used different parts of the body for measuring lengths. That's where units like the foot and the cubit came from.

It's not difficult to see how this way of measuring caused problems. The length of the cubit, for example, depended on the arm-length of the measurer! Cubits could, and did, have different lengths!

Standard Units

As long ago as 3000 B.C., the Egyptians solved the problem of how to measure lengths exactly. They did this by inventing the **standard cubit**. They realised that the length of the cubit didn't really matter as long as everyone used the same length. And so they marked out a cubit length on a piece of granite. Then they made measuring sticks exactly the same length as that standard cubit. In this way they made sure that the cubit was the same length all over Egypt.

That's really how measuring is carried out today. For each measurement, a **standard** is chosen. Every measuring instrument has to be compared with that standard:

Every 1 kg object must have the same mass as the standard kilogram, a piece of platinum kept in Paris. You can see the standard kilogram in the photograph on the right. It's the shiny metal cylinder inside the smallest jar. The air has been sucked out of the jars to make sure that the metal does not rust.

Every metre stick must be the same length as the standard metre. For many years, the standard metre was another piece of platinum kept in Paris, but now the standard metre is measured in a more complicated way, using a laser.

Every clock should keep the same time as an atomic clock. There are 30 special atomic clocks which are kept in different laboratories around the world.

The 'standard kilogram'

Did you know?

- Atomic clocks are carried by air from one laboratory to another for checking. This makes sure that times are accurate all round the world.
- An atomic clock only goes wrong by 1 second in 10 000 years.

1 a) Why were cubits and feet first used to measure lengths? ▲
   b) Why did this cause problems? ▲
2 How did the Egyptians make sure that every measuring stick was the same length? ▲
3 What is a standard used for? ▲
4 How would you check that your watch was keeping time?
5 Which animals still have their heights described as 'hands'?
6 **Try to find out:** what Weights and Measures Inspectors do.

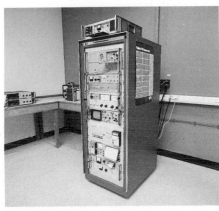

An atomic clock

1.2 Doing experiments

You will do lots of experiments in the science laboratory. In this unit, you will learn some tips which will help you to do these experiments well. Here is the first tip:

A good scientist always thinks carefully about what is happening in an experiment.

Two questions will help you to keep thinking. You should ask yourself these questions each time you do an experiment. They are:

'What happens in the experiment?'
'Why does this happen?'

Now look at these drawings of three 'separating' experiments. Try to work out what is happening in each experiment. Then answer the questions at the foot of the page.

1 Separating sand and gravel

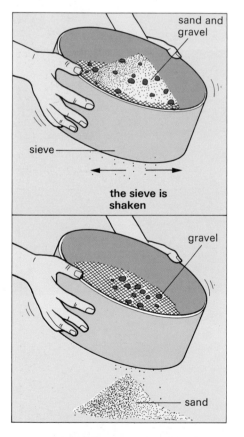

2 Separating iron and sand

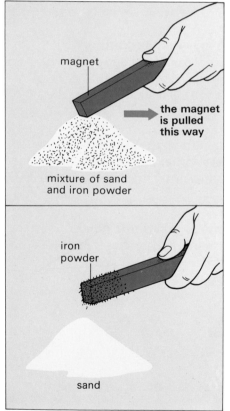

3 Separating chalk and water

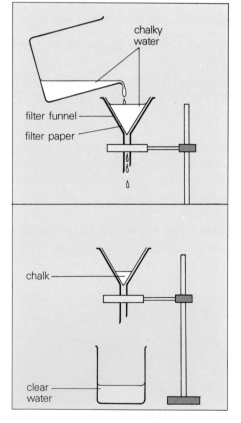

1 a) **What happens** when the sand and gravel are shaken in the sieve?
   b) **Why** does this happen?

2 a) **What happens** when the magnet is moved through the mixture of iron and sulphur?
   b) **Why** does this happen?

3 a) **What happens** when chalky water is put into the filter funnel?
   b) **Why** does this happen?

1.2 Safety first in experiments

A good scientist always works safely

It's very important to work safely. Carelessness can cause accidents.
That's why it is a good idea to have a set of safety rules to follow.

Use the pictures below to help you to write your own safety rules:

1 Spot seven differences between the two pictures.
2 Write a safety rule about each difference.

Science class AZ (a danger to everyone)

Science class AZ (on their best behaviour)

1.2 It helps to do some planning

A good scientist plans out each experiment carefully

Class AZ were in science. Their teacher was telling them about the experiment for the day. 'I want you to find out if the yellow flame from a bunsen burner is hotter than the blue flame,' she said. 'You can't do this by putting a thermometer in the flames – the bulb will crack. You get the answer by using the flames to boil some water.'

Ron and Stan both did the experiment, but they did it in very different ways.

Ron's rush

Ron wanted to be first with the answer. He rushed round the lab, grabbing the first apparatus he could find. He quickly ran water into two beakers (using both taps at the sink to speed things up). Then he lit the gas and started timing. Three minutes later he shouted out, 'The yellow flame's the winner.'

Ron's Rush

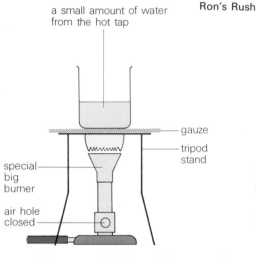

a small amount of water from the hot tap

gauze

tripod stand

special big burner

air hole closed

The water took 3 min to boil

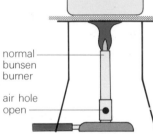

a lot of water from the cold tap

normal bunsen burner

air hole open

The water took 6 min to boil

Stan's plan

Stan spent five minutes thinking, 'How can I make a fair test?' Then he searched for the correct apparatus. He set it up carefully, lit the gas, and started timing. His results showed that the blue flame was the hotter.

The teacher didn't agree with Ron. She told him to think again and then repeat the experiment. But she was pleased with Stan. He had worked carefully and found the correct answer.

Stan's Plan

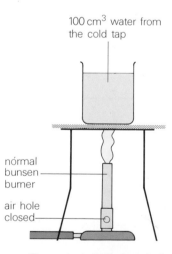

100 cm³ water from the cold tap

normal bunsen burner

air hole closed

The water took 7 min to boil

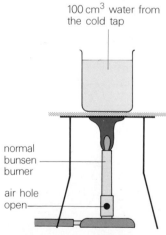

100 cm³ water from the cold tap

normal bunsen burner

air hole open

The water took 4 min to boil

1 Why did Ron decide that the yellow flame was hotter?
2 How did Stan 'make a fair test'?
3 Suggest three reasons why Ron got the wrong answer.
4 The owner of a car hire company wants to compare the distances which an Austin Metro and a Ford Fiesta can travel on 10 litres of petrol. Plan out a fair test he could use to do this.
5 **Try to find out:** some information about the petrol consumption of these cars.

Two questions to ask before you begin an experiment:

- 'What am I trying to find out in the experiment?'
- 'How can I find this out?'

1.2 It pays to keep your eyes open

For the enthusiast

A good scientist notes down everything he observes. This *includes* things which seem to be odd!

Many important scientific discoveries have been made after a sharp-eyed scientist has noticed – and noted down – something unusual. Here are three of these unusual observations, and the discoveries which followed them.

1 The observation In 1791, an Italian called Galvani was cutting up or **dissecting** a (dead!) frog's leg. He hung the leg from a copper hook. When he cut into it with an iron knife, the leg twitched.

and the discovery Galvani's friend, Volta, realised that electricity had made the leg twitch. He discovered that electricity had been produced when the two metals touched moisture in the frog's leg. In 1800 he used this discovery to make the first battery. It was made of silver and zinc discs separated by pieces of cloth soaked in water.

2 The observation In 1928, Sir Alexander Fleming (a Scot) noticed a fungus growing on a dish of bacteria. The fungus was killing the bacteria.

and the discovery It was later discovered that the fungus was producing a bacteria-killing chemical called *penicillin*. The first penicillin drugs were produced in 1943.

3 The observation In 1896, a Frenchman called Henri Becquerel, accidently left a photographic film in a drawer containing uranium crystals. Later he found an outline of the crystals on the film.

and the discovery Becquerel discovered that the uranium was giving off radioactivity. Radioactive uranium was later used in the first atomic bomb (1945) and the first nuclear power station (1956).

Did you know?

- The first nuclear power station was built at Windscale, Cumbria.
- Penicillin kills bacteria which cause pneumonia, tuberculosis and many other diseases.

1 What did: a) Sir Alexander Fleming b) Henri Becquerel observe? Why were these observations important? ▲
2 What made the frog's leg twitch when Galvani cut into it? ▲
3 You can build a battery from aluminium milk bottle tops, copper coins and wet paper. Draw a diagram to show how you would do this.
4 **Try to find out:** about other uses of radioactivity.

From this (1791)...
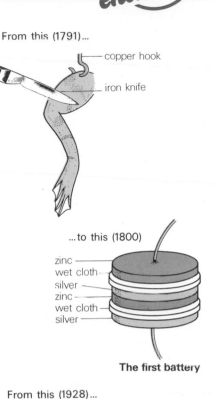
...to this (1800)

The first battery

From this (1928)...
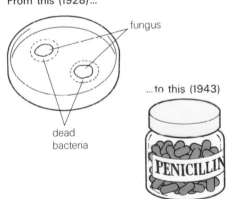
...to this (1943)

The first bacteria-killing drugs

From this (1896)...
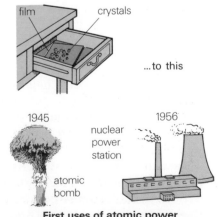
...to this

First uses of atomic power

9

1.3 Noticing differences

You don't have to be a first-class scientist to notice that people are different. Just look at the pupils in your science class! Some are tall, some are short. Some are thin, some are fat. Some have freckles, some don't. Some have brown hair, some blond, some black. There are different eye colours, too. The world is so much more interesting because people are so different.

The pupils in Science class AZ were very different from each other:

Joe can roll his tongue...

...but Angela can't! Can you?

pupil's name	height (cm)	shoe size	hair colour	eye colour	left- or right-handed	freckles or no freckles
Tom	153	4	blond	grey	right	no freckles
Anne	161	5	auburn	blue	right	freckles
Joe	150	2	brown	blue	right	no freckles
Joan	151	2	brown	green	left	freckles
Harry	151	5	blond	blue	right	freckles
Angela	150	4	black	brown	right	no freckles
Len	153	4	black	grey	left	freckles
Mary	162	6	brown	brown	right	no freckles
Ian	161	4	brown	blue	left	freckles
Jill	162	5	blond	blue	right	freckles
John	161	7	blond	green	right	no freckles
Diane	161	5	brown	blue	right	freckles
Bob	154	5	brown	brown	right	freckles
Louise	159	5	brown	blue	right	freckles
Stan	155	5	ginger	blue	left	no freckles
Gwen	151	3	brown	brown	right	freckles
Janet	154	4	auburn	green	left	freckles
Ron	158	6	black	brown	right	no freckles

1 Write down: a) Len's height b) Joan's eye colour c) Ian's hair colour. ▲
2 What is the most common: a) shoe size b) eye colour?
3 Describe: a) Bob b) Angela. ▲
4 Which pupil is most like you, and why?
5 How would this table of information have changed three years later when the class was 4 AZ?
6 Would the tallest-ever man have banged his head on your classroom door? Measure the door to find out.

Did you know?

- The tallest-ever human (a man) measured 272 cm.
- The shortest-ever human (a woman) measured 61 cm.

People can be very different!

1.3 Data in picture form

Scientific information is sometimes called **data**.

Scientists often collect lots of data. The table on the opposite page contains data about the pupils in class AZ.
Making a table is one way of writing down information. But a table containing lots of information can be quite confusing and quite dull! That's why scientists often put the data in picture or diagram form.

All of the information about class AZ can be put in picture or diagram form. You can use a **pie chart** or a **pictograph**, as shown on the right:

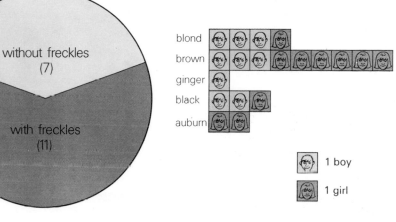

Pie chart showing the number of pupils in class AZ with and without freckles

without freckles (7)

with freckles (11)

Pictograph showing the hair colour of the pupils in class AZ

blond
brown
ginger
black
auburn

1 boy

1 girl

But a **bar chart** is the usual way of recording information:

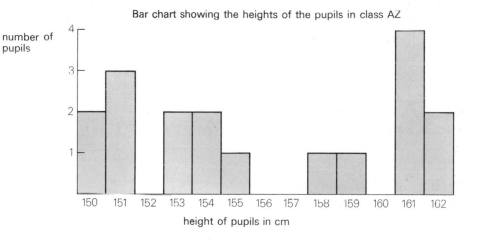

Bar chart showing the heights of the pupils in class AZ

number of pupils

height of pupils in cm

1 What is meant by 'data'? ▲
2 From the pictograph, work out the number of pupils with:
   a) blond hair   b) black hair.
3 From the bar chart work out:
   a) the number of pupils with height 151 cm
   b) the most common height and the number of pupils with that height
   c) the number of pupils with height 160 cm or more.
4 From the table of information on the opposite page:
   a) make a pictograph to show the pupils' eye colours. (Use ● to represent one pupil's eyes.)
   b) draw a bar chart of the pupils' shoe sizes.

Did you know?

● Freckles are affected by the sun. They often fade in winter because then the sun is less strong.
● The biggest shoe ever sold was size 42.

11

Looking at living things

Our world is full of factory-made things which are the same as each other.
The bottles in the picture are all exactly the same. That's the way the factory meant them to turn out.

But in the world of living things, there are lots of differences.

Sometimes, things which look different are really the same.

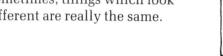

One man - two faces

One hare - two coats

hare in winter

hare in summer

But often there are real differences between one living thing and another.

What's special about living things? Are any living things like each other? Which differences really matter? That's what this section is all about.

Organism is the scientific name for a living thing. You are an organism living on the planet Earth.

Millions and millions of other organisms live on Earth with you. Very many of these belong to the group of **animals**. Many more belong to the group of **plants**.

Here are some of the things which animals are able to do:

Animals **move**

Animals **feed**

Animals **grow**

Animals **take in and get rid of gases**

Animals **react** to things about them

Animals **reproduce** (they produce others like themselves)

Many people think that plants are completely different from animals, but they are wrong. Plants also need **food** to live (but they can make it for themselves). Plants **grow** and **reproduce**, too. They **take in and get rid of gases** and they can **react** to things round about them. No plant can move around from place to place on its own, but parts of plants can **move**. Some plants open their flowers when the Sun shines, and others move the flower round to face the Sun.

Plants are more like animals than you would ever imagine!

1 Animals and plants are organisms. What does this mean? ▲
2 In what ways are animals and plants like each other? ▲
3 Make a set of animals and a set of plants from this list:
   cat, carrot, corn, eagle, whale, toadstool, seaweed, jellyfish.
4 How would you react if you sat on a drawing pin?
5 **Try to find out:** a) why plants grow upwards
   b) if there are signs of life on other planets.

Did you know?

On Earth there are:

- more than 1 500 000 different kinds of organism
- about 3 000 000 000 000 000 000 000 000 000 000 000 animals.

2.1 Animals can be very different

All sorts of different animals live on the Earth. Here are two animals you can easily study in the science laboratory. There are lots of differences between them.

The gerbil

Wild gerbils live in underground burrows in desert areas of India, China and North Africa. Gerbils make good pets. That's why they are often kept in science laboratories.

A gerbil has a **backbone** which supports its body. You can't see this backbone in the picture. It is inside the gerbil's body. But you can feel it when you stroke the gerbil.

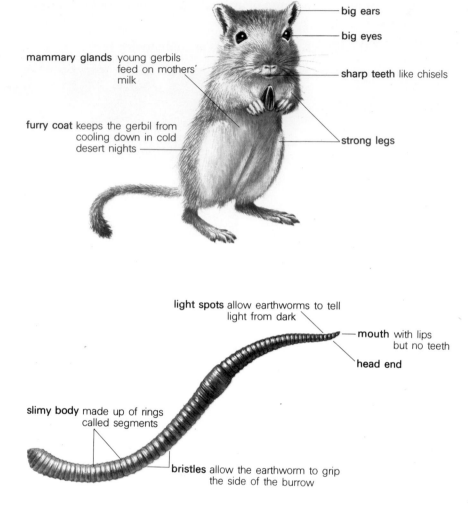

big ears

big eyes

sharp teeth like chisels

mammary glands young gerbils feed on mothers' milk

furry coat keeps the gerbil from cooling down in cold desert nights

strong legs

light spots allow earthworms to tell light from dark

mouth with lips but no teeth

head end

slimy body made up of rings called segments

bristles allow the earthworm to grip the side of the burrow

The earthworm

Earthworms are not usually kept as pets, but you don't have to go far to find one. They live in underground burrows in most parts of the world.
Earthworms have very bendy bodies. They don't have backbones. The earthworm's body is covered with slime called **mucus**. The mucus allows gases to pass into and out of the worm's body. It also allows the worm to slide through burrows more easily.

You can learn a lot by studying earthworms and gerbils in the science laboratory. But you would learn far more if you could study them in the place where they normally live, their **natural habitat**.

Did you know?

- The biggest earthworm ever measured was 6 m 70 cm long.
- A gerbil is a rodent. Rodents are animals with special teeth for gnawing their food.

1   What covers a gerbil's body? What covers an earthworm's body? In what way is each covering useful? ▲
2   Which animal (gerbil or earthworm) is better at spotting danger and escaping? Explain your answer.
3   What is meant by 'studying animals in their natural habitat'?
4   Why is it difficult to pull a worm out of its burrow? ▲
5   Would the biggest ever earthworm stretch across your classroom?
6   **Try to find out:** what is special about rodents' teeth.

If you want to see wild animals in their natural habitats, you really have to get out into the country. But, for now, you can imagine yourself on a day trip to the island of Great Craggie. There you will find plenty of animals to study.

As you can see from the drawing, there are several different habitats on the island. Woodlands, grasslands, hills, marshes, rivers and the lake are six of these habitats. Each has conditions which are different from all the others. In the hills, the ground is dry and stony, in the marsh it is wet and boggy, and so on.

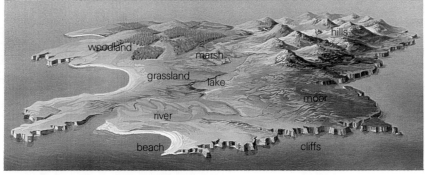

Animals live most of the time in the conditions which suit them best. Their bodies are better adapted to some conditions than to others. And so some animals are more likely to be found in one kind of habitat rather than in another.

soft damp skin (the moisture allows gases to pass in and out of the body)

webbed feet (for swimming)

strong leg muscles (for hopping)

tail (helps the squirrel to balance)

strong teeth

feet have 'fingers' and 'toes' for gripping

eyes and nostrils close under water

whiskers (feel vibrations in the water)

hind feet are webbed

Frogs are likely to be found in shady, damp conditions, near ponds, streams and ditches. They can live on land or in the water.

Squirrels are most likely to be found in woodlands. They live in the trees in nests called dreys and feed on nuts, seeds, cones and the tips of young twigs.

Otters live on the banks of rivers and ponds and on rocky coastlines. They feed on fish, crayfish and crabs.

1. Where are otters usually found, and why? ▲
2. Why must a frog stay in damp, shady conditions? ▲
3. Explain how:
   a) an otter's whiskers allow it to catch fish in muddy water
   b) a squirrel's feet and tail help it to move easily through the trees.
4. What are the six habitats mentioned in the passage? ▲
5. In which of these six habitats are you likely to see: a) a fieldmouse b) a heron c) a toad? Explain each of your choices.
6. a) Two other habitats are shown on the island. What are they?
   b) **Try to find out:** the names of some animals which live mainly in these two habitats.

Did you know?

- In winter, a frog hibernates at the bottom of a pond.
- A squirrel loses its coat twice a year. Its winter coat has softer, thicker fur than its summer coat.

15

2.2 Sorting animals into sets

Great Craggie is an island with lots of animals on it. The sea round about it has more animals still.

You can see a few of these animals in the pictures below:

A   B   C   D

E   F   G   H

I   J   K   L

M   N   O   P

It is often useful to put animals of the same type into groups called **sets**. Here are three different ways of doing this:

1 *You can make sets by thinking of where the animals live:*
   a) Put all the animals which live on the land in one set.
   b) Put all the animals which live in the water into another set.
   c) Which animals don't fit properly into either of these sets? Why?

2 *You can make sets by thinking of what the animals can do:*
   a) Make a set of animals which can fly.
   b) Pick an 'odd man out' from this set. Why did you choose it?

3 *You can make sets by thinking of what the animals are like:*
   Make sets of animals with: a) feathers   b) fur or hair   c) fins.

4 Which way makes the best sets of animals? Why?

(D E
  H
  N O)

A set of animals found in the stream

Did you know?

● Only three kinds of snake live in Britain: the grass snake, the smooth snake and the viper or adder.
● The adder is the only poisonous British snake.

16

2.2 Classification

Biologists are scientists who study living things. They have worked out a good way of putting animals into sets. They call it **classification**.

Biologists begin by dividing animals into two very big sets:

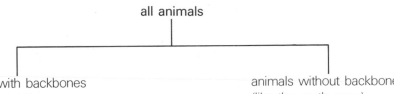

all animals

animals with backbones
(like the gerbil)

animals without backbones
(like the earthworm)

It's not easy to put the island's animals into these sets. You can't see whether an animal has a backbone just by looking at it, and you can't cut up live animals to find out! But you can get clues from the bodies of dead animals. It is also a help to remember that all animals with four limbs (arms, legs or wings) have backbones. (All the animals shown on the last page have backbones.)

Biologists divide animals with backbones into 5 more **subsets**:

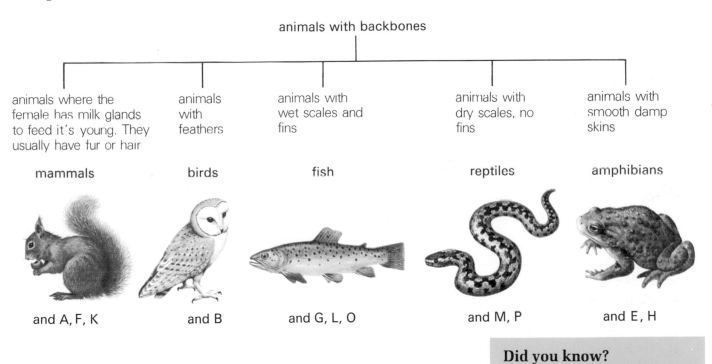

animals with backbones

| animals where the female has milk glands to feed it's young. They usually have fur or hair | animals with feathers | animals with wet scales and fins | animals with dry scales, no fins | animals with smooth damp skins |

mammals   birds   fish   reptiles   amphibians

and A, F, K   and B   and G, L, O   and M, P   and E, H

1. What is: a) a biologist b) a vertebrate c) classification? ▲
2. What is the difference between a fish and a reptile? ▲
3. Write down as much as you can about a mammal. ▲
4. a) Do all animals with four limbs have backbones? ▲
   b) Do all animals with backbones have four limbs? Explain.
5. Which set does each of these animals fit into:
   dog, tortoise, human, hen, newt, salmon, gerbil?
6. **Try to find out:** the names of some invertebrates.

Did you know?

- Animals with backbones are **vertebrates**. Animals without backbones are **invertebrates**.
- There are about 1 200 000 different kinds of invertebrate, but only about 47 000 different kinds of vertebrate.

2.2 Take a closer look!

It pays to look closely at an animal before you put it into a set. Appearances can be deceiving!

Here are two animals which you might find in the sea round Great Craggie. They both *look* like fishes, but one of them does not belong to the set of fishes.

Animal **G** is covered with very small scales, and it has fins. Its blood temperature changes with the temperature of the water round it. It stays under water without coming up for air. The young are not fed by their mother. Animal **G** is a **fish**. It is a **basking shark**.

But animal **F** has no scales or fins on its body (although, to be fair, you won't see hair or fur either). It's blood stays at a steady temperature. After 15 minutes under water, it has to come up for air. The young feed on their mothers' milk. And so animal **F**, a **dolphin**, is a **mammal**, not a fish.

Animal **K** is another which could confuse you, particularly as it only comes out at night. As it flies around it looks like a small bird. But it has fur, not feathers. Its blood temperature is constant. The females have milk glands. Animal **K** is a **bat**, the only mammal which can really fly.

1 Why is a shark put in the set of fishes? ▲
2 a) Why is a dolphin put in the same set as a bat? ▲
  b) How is the dolphin different from most other mammals?
  c) How is the bat different from most other mammals?
3 a) How is the blood temperature of a fish different from the blood temperature of a mammal?
  b) Why must Arctic fish produce 'anti-freeze chemicals' to survive in icy waters?
  c) Why can whales (mammals) survive without these chemicals?
4 A fish can use the oxygen which is dissolved in the water, but a dolphin can't. What difference does that make?
5 **Try to find out:** some things which dolphins have been trained to do.

Did you know?

- Next to man, the dolphin is considered to be the most intelligent creature.
- Arctic fish produce 'anti-freeze chemicals' to help them to survive in ice-cold waters.

18

2.3 Using keys

When you are trying to name animals, a special table called a **key** can be very useful.

Here is a key which will help you to name the island's mammals. To use it, you have to ask yourself some questions like 'does the animal have flippers?', 'does it have legs?' and 'does it have wings?'.

1 a) If the mammal has flippers, it is a **dolphin**.
   b) But if the mammal has legs, *you must go to* **2**.

2 a) If the mammal has wings, it is a **bat**.
   b) But if the mammal has no wings, *you must go to* **3**.

3 a) If the mammal has a bushy tail, it is a **squirrel**.
   b) But if the mammal has a long narrow tail, it is an **otter**.

The correct answers are: **A = otter   F = dolphin   I = squirrel   K = bat**

Now name the island's reptiles, fish and amphibians:

mammals

reptiles

fish

amphibians

Reptile with:
1 a) legs   →**lizard**.
    b) no legs   →*go to* **2**
2 a) zigzag pattern
        on skin   →**adder**.
    b) no zigzag pattern
        on skin   →**slow worm**.
    Name animals **P**, **C**, *and* **M**.

Fish with:
1 a) five gill slits
        showing   →**basking shark**.
    b) gills covered   →*go to* **2**
2 a) 3 sharp
        spines   →**stickleback**.
    b) no spines →*go to* **3**
3 a) round body→**trout**.
    b) flat body →**flounder**.
    Name animals **G**, **L**, **D**, *and* **O**.

Amphibians with:
1 a) tail   →**newt**.
    b) no tail   →*go to* **2**
2 a) smooth skin   →**frog**.
    b) warty skin   →**toad**.
Name animals **E**, **H**, *and* **N**.

Here are four of the island's invertebrates that you could find living under a log.

Make up a key to help someone to name the invertebrates.

earthworm   snail   spider   centipede

2.3 Identifying plants

In the drawings, you can see the flowers from six plants which grow on Great Craggie.

Here is a different kind of key. It will help you to identify the plants by their flowers.

Identifying plants by looking at their flowers can be useful. But there are snags:

- You can only identify the plants this way while they are in flower. And that may only be for a few weeks each year.
- You can't always rely on flower colours. Some plants can produce flowers of different colours.
- Some plants, like mosses, ferns and conifers (trees which produce cones) don't produce flowers at all.

Fortunately, there are other clues which you can use. The plant's leaf, stem and fruit can be a help. You can even get a clue from the place in which the plant is growing. The bluebell grows best in woodlands, in rich soil and sheltered conditions. The sundew grows in very wet, boggy areas. And so a blue flower at the top of a mountain is unlikely to be a bluebell. A white flower growing on dry sandy soil won't be a sundew.

Great Craggie flowering plants

- large 'flower' made up of many small flowers packed tightly together **clover**
- single flower
  - separate petals
    - all petals the same shape
      - 4 petals on flowers **charlock**
      - 5 petals on flowers **sundew**
    - petals have different shapes **vetch**
  - joined petals
    - flowers grow evenly round the stalk **bell heather**
    - flowers grow on only one side of the stalk **bluebell**

leaflets

leaf

leaflets

leaf

The leaves form a ring at the bottom of the stem. This is a **rosette**

the leaves have hairs

whorls

1 Why can't you always identify plants from their flowers? ▲
2 Is sandy soil suitable for growing sundew? Explain. ▲
3 Use the flower key to answer these questions:
   a) What does a charlock flower look like?
   b) In what way is clover the 'odd man out'?
   c) What are the names of flowers **A** to **F**?
4 Look at the plants' leaves. Which plant (or plants) has leaves which: a) form rosettes b) have leaflets c) are in whorls d) have smooth edges e) have jagged edges?
5 Make a key to identify the plants from their leaves.

2.3 Scientific sets

In this chapter, you have only been able to look at a small part of the Great Craggie wild life. But as you will probably have guessed, there are many other animals and plants living on the island.

In particular, there are far more birds than those shown on page 16. Many birds, like the tawny owl, live on the island all year round. They are called **residents**. Others come only in winter (like the redwing) or in summer (like the skylark). A small number will take a few days' rest on the island as they **migrate** from one place to another.

Species and genus

Birds of the same kind are put into a set called a **species**. The tawny owls are put into one species. The skylarks are put into another. In a year, 100 or more different species of bird may be seen on the island.

You would really need a bird book with a good key to identify all these birds. Some bird species are closely related to each other, and look alike. Bird species which are closely related are put in a set called a **genus**. The gulls in the pictures belong to one genus, swans to another, falcons to yet another.

In a bird book, you will see that each bird has a scientific name made up of two parts. In fact, all living things have been given two-part names. The first part, which always begins with a capital letter, gives the name of the genus. It's a bit like your surname. The second name, which always begins with a small letter, is the species name. It tells you exactly which member of the genus is being described. It's a bit like your Christian name. Here are the scientific names of four of the gull family:

common name	scientific name
herring gull	Larus argentatus
lesser black-backed gull	Larus fuscus
little gull	Larus minutus
black-headed gull	Larus ridibundus

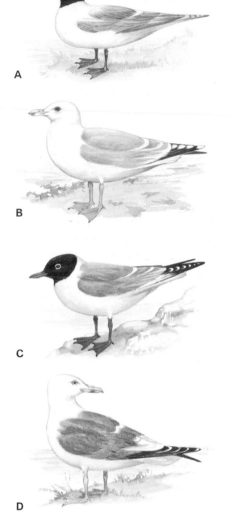

A

B

C

D

These are pictures of the four gulls mentioned in the table. Which gull is which?
(See question 3b)

1 What is meant by: a) a genus b) a species c) a resident? ▲
2 Why is it often difficult to identify birds in the same genus? ▲
3 a) In the table, you can see the scientific names for the herring gull, the black-headed gull, the little gull and the lesser black-backed gull. What is the genus name of these gulls? Explain your answer.
   b) These four gulls are shown in the pictures. Use a bird book to decide which gull is which.
4 Explain why using a scientific name can be less confusing than using a common name.
5 **Try to find out:** a) the scientific name of the tawny owl, the redwing and the skylark b) more about bird migration.

Did you know?

- There may be lots of common names for the same animal or plant. The creeping buttercup has 36 different names in Britain alone.
- The scientific name of an animal or plant is the same all over the world.

Energy

You often hear people talking about energy:

'Where do they get their energy from?'   That's the kind of thing which tired parents – and grandparents – often say about young children.

'Save energy!'   That's a message which Governments have been trying to get over for a number of years.

'Energy crisis ahead'   That's the kind of warning you can see in newspaper headlines or on television.

It's not surprising that there is so much 'energy talk'. Energy is needed:

- to keep your body working
- for heating homes, schools and offices
- to drive machines in industry
- to power cars, buses, aeroplanes, ships and other forms of transport.

In fact, our whole way of life depends on good supplies of energy.

That's why this section is all about energy. It looks at different kinds of energy, and at ways of changing one kind of energy into another. It also looks at the world's energy supplies, at supplies which are running out and at new energy supplies for the future.

You could say that energy keeps the whole world running!

All sorts of energy are needed to make an exciting football match!
In the picture below, you can see five kinds of 'energy in action':

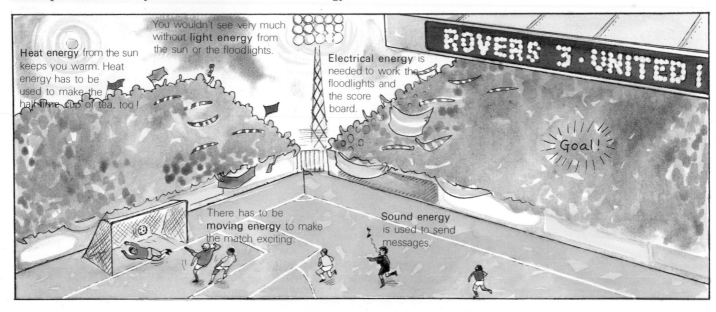

Heat energy from the sun keeps you warm. Heat energy has to be used to make the half-time cup of tea, too!

You wouldn't see very much without light energy from the sun or the floodlights.

Electrical energy is needed to work the floodlights and the score board.

ROVERS 3 · UNITED 1

Goal!

There has to be moving energy to make the match exciting.

Sound energy is used to send messages.

Storing energy

Energy can also be stored up, ready for action. You can store up energy in a bow by pulling the bow string. (When you let the string go, there is plenty of action!) The stretched bow has **stored energy**. So does water held back by a dam. So does a sledge at the top of a hill.

The battery in your watch, or radio is a store of **chemical energy**. So is the food you eat. So is a fuel like coal or petrol.

◄ How many 'energy stores' can you see in this picture ?

Uranium is the fuel used in nuclear power stations and in 'atomic' bombs. The energy stored in it is called **nuclear** or **atomic energy**.

1 Write down: a) five kinds of energy in action   b) three stores of chemical energy   c) three other things which have energy stored in them. ▲

2 Which store of chemical energy: a) keeps the referee's watch working   b) gives the players their energy   c) keeps the team bus running?

3 Write down four things which: a) use electrical energy
   b) produce heat energy   c) have moving energy
   d) produce sound energy.

4 **Try to find out:** a) how energy is stored in a 'wind up' watch
   b) what is meant by a 'potential' international player.

Did you know?

- Moving energy is also called **kinetic** energy.
- A 'moving picture' at the cinema used to be called a 'kinematograph'.
- The stored energy in the stretched bow string, in the water behind the dam, and in the sledge at the top of the hill is also called **potential** energy.

3.1 Energy changes

Over one hundred years ago, scientists did some careful experiments on energy. This is what they found out:

Energy *cannot* be made or destroyed,
but **energy *can* be changed from one form to another.**

You will be able to understand this better if you think about a light bulb. The bulb does not make light on its own. (It does not shine when it is in the packet!) But it does shine when electricity is flowing through it. Then it changes the electrical energy to light energy.

The main energy change in the light bulb is written like this:

electrical energy ⟶ light energy
(the arrow means 'changes to')

In a light bulb:
electrical energy ⟶ light energy

Here are some other main energy changes:

In a coal fire:
chemical energy (stored in coal) ⟶ heat energy

For a diver:
stored energy ⟶ moving energy

In an electric truck:
chemical energy (stored in battery) ⟶ electrical energy (in wires) ⟶ moving energy (from motor)

1 What did scientists find out about energy around 100 years ago? ▲
2 An electric light bulb cannot make light out of nothing. What does happen in the bulb when it shines? ▲
3 What are the energy changes: a) when you play a guitar b) when you ski downhill c) when you strike a match?
4 What energy changes take place in: a) a torch b) a firework c) a car engine?
5 Look at the cartoon strip below. Four kinds of energy appear in it (at **A**, **B**, **C** and **D**). What are they?

Did you know?

- Most modern light bulbs last for 1000 hours.
- One light bulb in a fire station in California has been shining since 1901.

3.1 Energy changers

1 Energy changers in the kitchen

Find eight energy changers in the picture.
Then make a table to show each energy changer with its energy change:

energy changer	energy change
Kettle	electrical → heat

2 Muddled-up energy changers

The boxes on the right give the muddled-up names of five energy changers. Sort out each name. *Then write it down and answer the question about it:*

1   You probably meet this energy changer first thing in the morning.
   a) What kind of energy does it produce?
   b) What kinds of energy can be used to make it work?
2   This energy changer is often used after swimming. What energy changes take place in it?
3   This can give music wherever you go. What store of energy makes this possible?
4   The average British family uses this energy changer many hours each week, mostly in the evening. What kinds of energy does it produce?
5   This is intended to produce moving energy. It has wheels.
   a) What store of energy does it carry?
   b) Which kind of energy lets you know it is coming?

Did you know?

- The cooker uses more energy than all of the seven other kitchen energy changers put together.
- A fluorescent 'strip' light only changes 20% of the electrical energy to light energy.

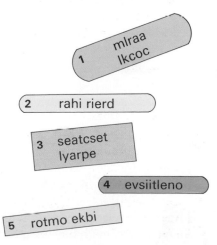

1   mlraa lkcoc

2   rahi rierd

3   seatcset lyarpe

4   evsiitleno

5   rotmo ekbi

3.1 Energy changers change the world

Electric motors and **electric generators**, or **dynamos**, are among the most important energy changers which have ever been invented.

You can make a motor and a generator from a coil of wire. The coil has to be fixed so that it can spin between the poles of a magnet:

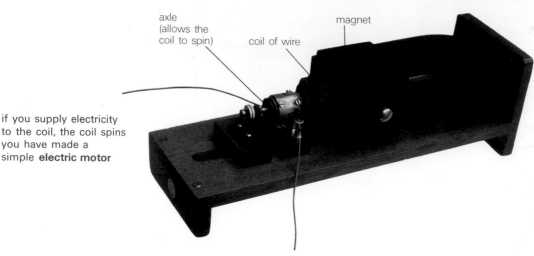

axle (allows the coil to spin)

coil of wire

magnet

if you supply electricity to the coil, the coil spins you have made a simple **electric motor**

if you use a steam engine or a water wheel to turn the coil, electricity is produced in the coil
This time you have made a simple **electric generator** or a **dynamo**

Electric motors in everyday use are built with several coils of wire. This makes them more powerful and smoother running. There are electric motors inside electric drills, electric cars, vacuum cleaners and most other machines which change electricity to movement.

Some bicycle dynamos are built with only one coil of wire. But the power station generators, which produce our household electricity, are much more complicated.

Inside the locomotive of an electric train is an unusual 'motor-generator unit'. It can act both as a motor and as a generator. When electricity is supplied to the unit, it acts as a motor. The coils spin and this movement drives the wheels. When the power is switched off, and the train slows down, the unit acts like a generator. The coils spin as long as the wheels are turning. This produces electricity which can be used for heating and lighting the train.

Did you know?

- The first really useful electric motor was built in 1873.
- Only six years later, the first electric train was built.

1 What would you need to make an electric generator? ▲
2 Name six machines which contain electric motors.
3 What is the energy change in: a) an electric motor b) a dynamo?
4 What is happening in the 'motor generator unit' of an electric train when: a) the train is speeding up b) the train is slowing down?
5 The energy changers opposite have also 'changed the world'. Suggest why each has been so important.

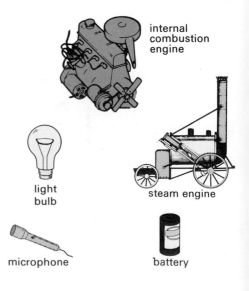

internal combustion engine

light bulb

steam engine

microphone

battery

3.2   Energy for your body

The question 'What is your body's energy source?' really means 'Where does your body get its energy from?'.

Your body has to get energy from somewhere   It uses up energy all the time. Your body uses up energy:

- when it grows
- when your muscles move
- to keep you warm

Your body gets the energy it needs from food   There are large amounts of chemical energy stored in energy-rich foods like bread, cakes, cornflakes, crisps, chips and sweets. You can make use of this energy by eating these foods. Your body changes the chemical energy to the kinds of energy it needs, like heat energy and moving energy.

The energy in food comes from the Sun!   Green plants use sunlight to grow. They change the light energy into useful chemical energy.

Many plants build up stores of chemical energy which are then used to make your food. Wheat grains, for example, are chemical energy stores built up by the wheat plant. Wheat grains are used to make the flour for baking bread and cakes.

All these facts point to one thing:

the Sun is your body's energy source.

The energy you use on a wet winter run comes from summer sunshine!

Foods for energy

light energy

5 miles later -still going strong

1   Why must your body be supplied with energy? ▲
2   Your body is an energy changer. Write down: a) the kind of energy it uses up   b) two kinds of energy it produces. ▲
3   Which energy change is carried out by green plants? ▲
4   Make a list of energy rich foods you have eaten today.
5   Explain why 'the energy you use on a wet winter run comes from summer sunshine'.

Did you know?

- There is enough energy stored in one large slice of toast to keep a cross country runner going for about 10 minutes.
- Plants use less than 1/100th of the Sun's energy which reaches Earth.

3.2 Too much and too little to eat

The amount of energy which your body needs each day depends on a number of different things. It depends on your age, your size, whether you are a boy or a girl, how active you are, and other things, too. It's impossible to calculate *exactly* how much 'energy food' you should eat in a day, but that does not matter. As long as you eat roughly the correct amount, you should have no energy problems.

The people who *do* have energy problems are those who eat far too much or far too little energy food. George and Rajan have energy problems. But their problems are very different!

The lists on the right show the food which George and Rajan could eat in a normal day. The energy contained in the food is given too. It is measured in **kilojoules (kJ)**:

George:
- lives in England
- works in an office
- drives to work
- reads, and watches television in his spare time

Rajan:
- lives in South India
- works on his small 'farm'
- works all day, and has little spare time

George's food	energy supplied
cereal (1 bowl)	325 kJ
bacon (2 rashers)	1430 kJ
eggs (2)	660 kJ
steak	960 kJ
potatoes (4)	1600 kJ
apple pie	2000 kJ
custard	280 kJ
sausages (4)	2300 kJ
chips	900 kJ
beans	460 kJ
bread (6 slices)	2000 kJ
butter (on bread)	1000 kJ
cake (2 slices)	1350 kJ
sugar (in tea)	800 kJ
biscuits (3)	1200 kJ
chocolate (1 bar)	2500 kJ

Rajan's food	energy supplied
rice (2 bowls)	2350 kJ
potato (1)	400 kJ
beans (1 bowl)	1650 kJ
other vegetables	450 kJ
banana (1)	600 kJ
sugar (2 spoons)	470 kJ
milk	370 kJ
coconut oil	900 kJ

1 What unit is used to measure food energy? ▲
2 Different people should eat different amounts of energy food. Explain why. ▲
3 Men normally need between 11 000 and 15 000 kJ of energy in a day. Office workers need around 11 000 kJ. Labourers need around 15 000 kJ. Explain the difference.
4 Add up the number of kJ in: a) George's food b) Rajan's food.
5 Energy food which is not used up is changed to fat.
   a) Explain why George is so fat.
   b) Which foods should he cut out if he wants to slim?
   c) What else could he do to lose weight?
6 Rajan is often too tired to work hard. Explain why.
7 **Try to find out:** why so many people in the world are short of food and what can be done about it.

Did you know?

- ⅛ of the world's population is starving.
- ⅓ of the people in the world do not have enough to eat.

Most of the world's energy comes from five main sources. The pie chart opposite shows how much energy comes from each source.

You may wonder why electricity is not one of the energy sources. After all, large amounts of electricity are used in homes and industry. The reason is that:

***all* electrical energy is produced from other sources of energy.**

Some electricity—**hydroelectricity**—is produced from the stored energy of water held back by a dam. The water runs downhill from the dam, through a pipeline, to the power station. There it turns huge motors called **water turbines**. The turbines turn generators, and the generators produce electricity.

Some countries, like Norway and Sweden, produce large amounts of hydroelectricity. Most countries, however, make their electricity using the chemical energy stored in coal, oil and gas, and the nuclear energy in uranium. The fuels are used to heat water in huge boilers, producing steam. This steam is used to turn **steam turbines**, and they turn the generators.

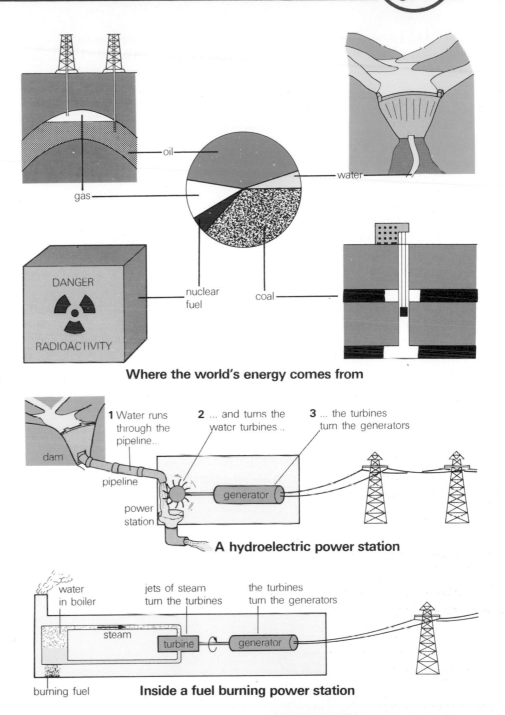

Where the world's energy comes from

A hydroelectric power station

Inside a fuel burning power station

1 Use the pie chart to name: a) the world's five main energy sources   b) the source which supplies most energy. ▲
2 Why is electricity not called an energy source? ▲
3 How is coal used to make electricity? List the energy changes. ▲
4 a) Give the energy changes when hydroelectricity is made.
   b) Will the energy source for hydroelectricity ever run out? Explain.
5 **Try to find out:** a) the name of your nearest power station and the source of energy it uses   b) why Norway can produce so much hydroelectricity.

Did you know?

● The jet of steam which turns a turbine travels at around 400 metres per second.
● 1 kilogramme of uranium can produce as much energy as 60 tonnes of coal.

29

3.3 When the fuels run out

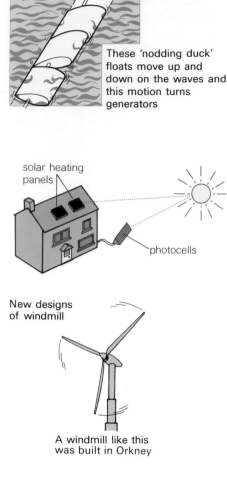

Going further

In the 1980's, most of the world's energy will be supplied by fuels – coal, oil, gas and uranium. But these fuels won't last forever. If most experts are right, some fuels will run out in your lifetime.

The bar chart opposite shows you how long present supplies of fuel are expected to last. Of course, the chart could change. If everyone decided to use fuels more sensibly, they would last longer.

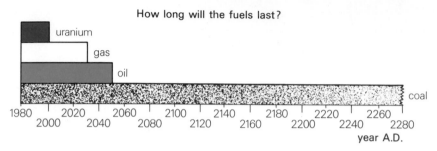

How long will the fuels last?

uranium
gas
oil
coal

1980 2000 2020 2040 2060 2080 2100 2120 2140 2160 2180 2200 2220 2240 2260 2280

year A.D.

These 'nodding duck' floats move up and down on the waves and this motion turns generators

When some of the fuels run out, there could be a serious shortage. That is why scientists and engineers are working hard to find ways of getting useful energy from new energy sources. They are trying to find ways:

* to use the moving energy of the wind and waves to turn generators to make electricity
* to use the sun's heat energy to heat homes, and to make cheap photocells to change its light energy to electricity
* to use the heat energy of 'hot rocks' which lie just below the Earth's surface
* to control the tremendous energy from nuclear fusion reactions which produce temperatures as hot as the sun
* to produce new fuels – like hydrogen from water or alcohol from sugar cane or other plants.

By the time the fuels run out, at least *some* of these problems should have been solved!

solar heating panels

photocells

New designs of windmill

A windmill like this was built in Orkney

This is a French design of windmill

Did you know?

* A nuclear fusion reaction produces a temperature of 100 000 000 °C.
* Plans have been made to build windmills as tall as skyscrapers, with blades 50 m long.

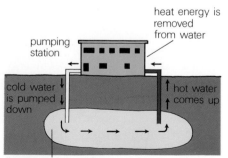

heat energy is removed from water

pumping station

cold water is pumped down

hot water comes up

the hot rock is broken up so that water can flow through

1 Why could there be a serious energy crisis in the future?
2 When are the following fuels expected to run out:
  a) oil b) gas c) uranium d) coal? ▲
3 What does a photocell do? ▲
4 Which energy sources will still be available when you are 70?
5 Why would it be very useful to be able to make: a) hydrogen from water b) alcohol from sugar cane?
6 **Try to find out:** some ways of saving fuel.

3.3 An energy trail

As you go round this energy trail, you will learn something about Britain's energy supplies in the next 50 years or so. You can use dice and make the trail into a game if you like. If not, jump round the trail two squares at a time. When you land on a red or green square, do what you're told!

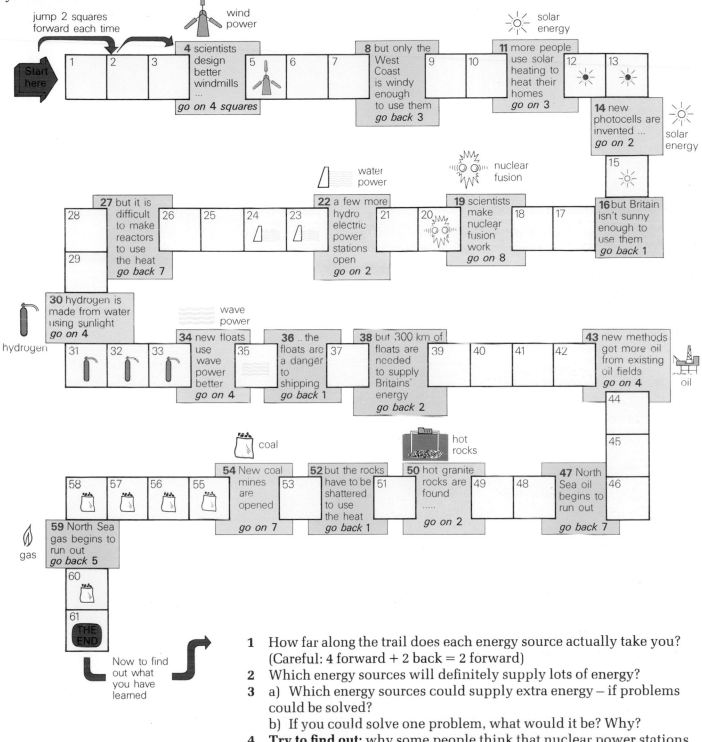

1 How far along the trail does each energy source actually take you? (Careful: 4 forward + 2 back = 2 forward)

2 Which energy sources will definitely supply lots of energy?

3 a) Which energy sources could supply extra energy – if problems could be solved?
b) If you could solve one problem, what would it be? Why?

4 **Try to find out:** why some people think that nuclear power stations are a big step forward, but others think exactly the opposite.

31

Building blocks

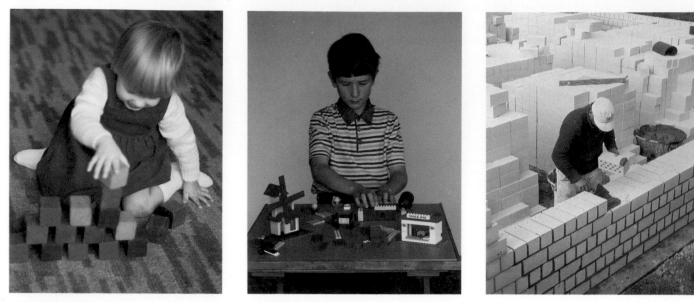

4

Toy building blocks, model building blocks and real building blocks are all easy to handle, if you know how! But dealing with nature's building blocks is much more difficult. That's because they are so very, very small.

It's difficult to realise just how small nature's building blocks are, but the model in the photo will help. It is a model of the chemical building block which makes up rubber. A model of a rubber band, built the same way, would stretch for many kilometres!

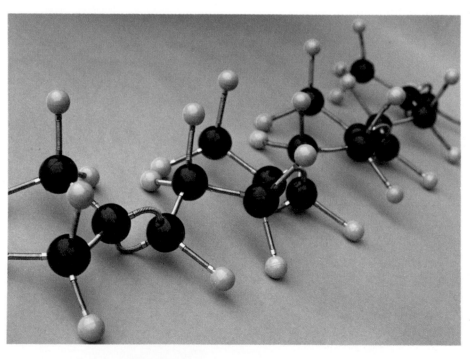

Scientists call nature's building blocks **atoms** and **molecules**. Finding out about them wasn't easy, and it did take a long time. But now scientists do know:

- what atoms and molecules are
- how atoms and molecules are arranged
- what atoms and molecules can do.

And so should you – after you read this section!

4.1 The three states of matter

Coal, heating oil and butane are three common fuels. They will help you to understand the differences between solids, liquids and gases.

The solid state
Coal is a **solid** fuel.
- A solid cannot move. It stays where it is, unless something, or someone, moves it.
- A solid keeps its shape— unless it is broken.
- The volume of a solid always stays the same.

The liquid state
Heating oil is a **liquid** fuel.
- A liquid can flow.
- A liquid takes up the shape of its container.
- The volume of a liquid stays the same.

The gas state
Butane is a **gas** fuel. It is the main fuel used for camping stoves and cigarette lighters.
- A gas can flow: it will spread out as far as it can.
- A gas can change shape.
- The volume of a gas can change, too.

It's difficult to show all of this. Most gases are invisible.

It's a solid, a liquid or a gas!

All of the substances around you can be put into the set of solids, or the set of liquids, or the set of gases. Solids, liquids and gases are **the three states of matter**.

1   What are the three states of matter? ▲
2   What can you say about: a) the shape of a solid   b) the volume of a liquid   c) the shape of a gas   d) the volume of a solid? ▲
3   Make a list of six solids, six liquids and six gases.
4   If North Sea gas is invisible, how can you tell when gas is leaking?
5   **Try to find out:** the names of some other fuels and whether they are solids, liquids or gases.

Did you know?
- About 30 million tonnes of fuel oil are used in Britain in a year.
- 40% of that oil is used by electricity power stations.

33

4.1 Different states – different jobs

Solids, liquids and gases are useful for different jobs. That's because they behave so differently.

Just think of how a car is designed:

glass — steel — air — petrol — plastic

Solids are used to make the car body. (The body has to be solid to keep its shape)

Liquids are mostly used as fuels. (The liquid fuels flow easily from the tank to the engine)

Gas is used to fill the tyres (A tyre with gas in it can change shape when it hits a bump)

It can be useful, too, to change a substance from one state to another – for example, from solid to liquid, or from liquid to gas:

A car engine is made of iron. But it cannot be shaped from a solid lump. Instead, the iron is melted and the red hot liquid is poured into a mould. The liquid flows into shape. Then it cools and hardens.

The glass for the windows is made in the same kind of way. Red hot liquid glass is allowed to flow until it forms a sheet. Then it cools and hardens.

Some cars use petroleum gas as a fuel. It would take up far too much space to store the fuel as a gas. And so it is stored in strong tanks as a liquid. This is changed to a gas when needed.

Did you know?

- In the 1940s some cars used gas fuel. They carried huge gas bags on the roof.
- Iron has to be heated to 1539 °C before it melts.

1 Explain why: a) a car body is mostly made of solids b) liquids are usually used as car fuel c) gas is used to fill car tyres. ▲
2 Why do cars not use: a) solid fuel b) solid tyres?
3 Why is it useful: a) to make a car engine out of red hot liquid iron b) to store petroleum gas as a liquid? ▲
4 Give some examples of:
   a) solids which are shaped by pouring liquid into a mould
   b) gases which are stored as liquids in cylinders.
5 **Try to find out:** a) how to make a plaster cast of a footprint
   b) how bronze sculptures are made.

The wartime petrol shortage wasn't a problem for the owner of this car?

4.1 Mind what you say

The words 'vapour' and 'fluid' are used to describe lots of everyday substances. But in science, both words have special meanings:

A *vapour* is a gas given off by a liquid, even when the liquid is not boiling.

At a filling station, the air is full of petrol vapour. You can smell it, but you can't see it. It is a colourless gas.

When water boils, a special sort of water vapour called **steam** is given off. But the white clouds which are produced are not steam. They are made up of tiny drops of liquid water. The steam is the invisible gas next to the surface of the boiling water.

A substance can be described as a *fluid* when each part of it – no matter how small – can change shape easily.

Liquids and gases are fluids. Sugar syrup is a fluid – when you pour it into a bowl, it flows and *exactly* takes up the shape of the bowl.
Sugar can also be poured, but it is *not* a fluid. It is made up of tiny crystals which can quite easily be seen using a hand lens. Each crystal is hard, and cannot change shape. Sugar is a **powder**, a crushed-up solid.

these clouds are droplets of **water**

steam is invisible

fuel vapour escapes whenever a car is being filled up

syrup pouring

sugar piled up

1 Does a vapour: a) keep its shape b) keep its volume c) flow?
2 What is meant by: a) a fluid b) a powder? ▲
3 Sugar is easily poured. Why is it not a fluid? ▲
4 From the list given below pick out: a) a set of fluids b) a set of powders c) a set of liquids which give off vapours you can smell, but not see:
air, perfume, sand, paint, flour, glue, salt, North Sea gas.
5 What is 'fluidised coal' used for? Is it a fluid? Explain your answer. ▲
6 Powdery ice often collects round the 'ice box' in the fridge. Where has it come from?
7 **Try to find out:** what a vapour trail is. Is 'vapour' the correct word to describe it?

Did you know?

- Some modern ships use powdered coal as fuel. It is called 'fluidised coal'. Air is used to blow it along pipes from the storage bunkers to the boilers.
- Syrup is 75% sugar.

4.2 What is everything made of?

Party balloons always go down sooner or later. The air inside always escapes.

It's easy to see how the air escapes if you stick a pin in a balloon, or if you don't tie it properly. But air can escape even from a tightly tied balloon, and that's more puzzling.

How does the air escape?

Here's an explanation that scientists agree on:

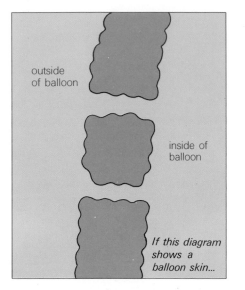

outside of balloon

inside of balloon

If this diagram shows a balloon skin...

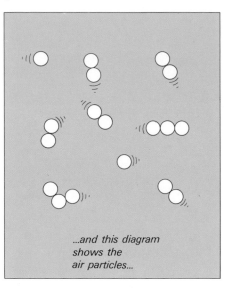

...and this diagram shows the air particles...

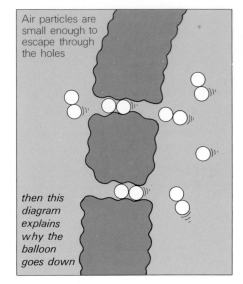

Air particles are small enough to escape through the holes

then this diagram explains why the balloon goes down

1 Balloon rubber behaves as if it has very small 'holes' in it.

2 Air is made up of millions of tiny particles moving about.

3 The air very slowly escapes through the 'holes'.

You can't really *prove* that this explanation is correct. You can't see particles escaping from a balloon, even with the most powerful microscope. But scientists are quite sure that balloon rubber has tiny holes in it. They are also sure that air is made up of tiny moving particles.

In fact scientists are sure that solids, liquids and gases are *all* made up of tiny particles called **atoms** and **molecules**. Atoms are the smallest particles. A molecule is a group of atoms joined together.

atoms join to give:

a molecule

1. Why does a balloon go down quickly if you stick a pin in it?
2. Air slowly escapes from a balloon even when it is tightly tied. Explain why. ▲
3. Write out and finish this sentence: 'Solids, liquids and gases are all made up of . . . .' ▲
4. Write down three pieces of information (from this page) which suggest that atoms are tiny.
5. What is a molecule? ▲

Did you know?

- More than 1 000 000 000 atoms are needed for the ink to make this dot →·
- Some molecules are made up of more than 500 atoms joined together.

4.2 More about molecules

You can learn more about molecules from 'mixing' experiments:

Diffusion

 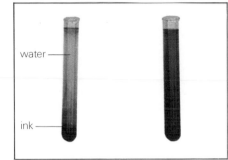

1 Stink bomb gas quickly spreads through the air.

2 Ink spreads through water, but it spreads more slowly.

3 When a dye crystal dissolves in water, the colour slowly spreads.

This spreading of one substance through another is called **diffusion**. When two gases (or two liquids) mix, they spread easily through each other. This suggests that: **the molecules in liquids and gases are moving.**

The 'missing' volume

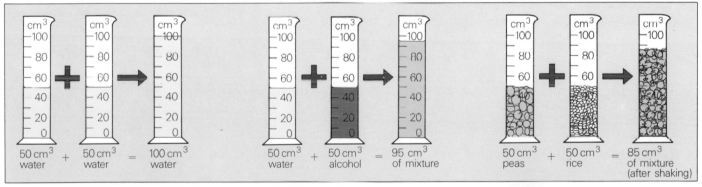

When you mix alcohol and water, the volume seems to shrink. You can explain this 'disappearing volume' by saying that:

the molecules in a liquid have spaces between them.

When the alcohol and water are mixed, water molecules fit into some of the spaces between the alcohol molecules. This makes the volume less than you would expect. If you look at what happens when you mix peas and rice, you can see that this makes sense.

The rice fits into the spaces between the peas. This makes the volume less

1 What is meant by diffusion? ▲
2 Why do two gases spread easily through each other?
3 Do molecules move faster in gases or liquids? Explain your answer. (The diffusion experiments will help you with this.)
4 Adding 50 cm³ alcohol to 50 cm³ water gives a volume of only 95 cm³. Explain why. ▲
5 a) How far could gas molecules travel in 1 second? ▲
  b) Why does stink bomb gas take so long to cross a room?

Did you know?

● Gas molecules move with average speeds of between 100 and 2000 metres per second. But they only travel tiny distances before they bump into other molecules.

The experiments on the last two pages don't prove anything about atoms and molecules. But experiments like these did help scientists to get a picture of the atoms and molecules in solids, liquids and gases.

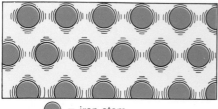

 = iron atom

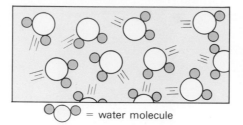

= water molecule

= oxygen molecule

This is a model of the atoms in a lump of **solid** iron.
Notice that the atoms are:
1 arranged in rows
2 close together
3 held together tightly
4 not able to change places
5 vibrating (moving backwards and forwards)

This is a model of the molecules in a drop of **liquid** water.
Notice that the molecules are:
1 not arranged in any particular way
2 close together
3 not so tightly held together as in solids
4 moving about and changing places

This is a model of the molecules in oxygen **gas**.
Notice that the molecules are:
1 not arranged in any particular way
2 far apart
3 very weakly held together
4 moving very fast in all directions

Scientists think that these are good models for all solids, liquids and gases. You can use the models to explain lots of things, like:

Why solids *can't* flow:

In ice cubes, the molecules are fixed in rows. The cubes can't flow.

Water flows because its molecules can change places.

Why solids *don't* split easily:

In ice, the molecules are more tightly held together than in water...

Why gases *can* be squashed easily:

Air can be easily squashed. Air molecules have lots of space between them

Iron can't be easily squashed. Its atoms are close together

1 Which of the following phrases describe atoms and molecules in
   a) a solid   b) a liquid   c) a gas
   close together   changing places   tightly held together   far apart
2 Why do: a) ice cubes not flow b) beach balls squash easily? ▲
3 Why do: a) gases spread quickly b) solids keep their shapes?
4 A guitar string *vibrates* when you pluck it. Atoms in a solid vibrate, too. In what way are they moving?
5 **Try to find out:** the name of the hardest solid.

Did you know?
- Atoms and molecules move faster when they are heated.
- More than 99.99% of the volume taken up by a gas is empty space.

4.2 Building up a theory

The Greeks first thought up the idea of atoms as long ago as 400 B.C. They thought that every substance was made up of tiny particles, and they called these particles atoms. They decided that atoms of different substances were very different from each other. They thought (wrongly!) that atoms of iron were hard and had hooks which fixed them together in a solid, that 'water atoms' were smooth and slippery and that 'air atoms' whirled about everywhere.

The Greeks were only guessing, but their guesses were very sensible. The trouble was that they did not *test out* their ideas. Certainly, the Greek ideas were better than the ideas 'scientists' had for the next 2000 years. Atoms went out of fashion. 'Scientists' thought that every substance was made up of earth, air, water and fire mixed together!

Scientists started to think about atoms again around 1650. In 1808, a Lancashire schoolteacher called John Dalton did some careful experiments with chemicals. His results convinced scientists that the 'atoms idea' was probably correct. As more and more experiments were done, more and more evidence piled up, making the atoms and molecules model clearer and clearer. The guesses became a **theory** – an idea which had been tested and found to fit a broad variety of facts.

Sometimes it took years to fit the results of an experiment into a theory. But often the experiments which were most difficult to explain helped to form the best theories.

For hundreds of years, 'scientists' called **alchemists** looked for the 'philosopher's stone'. They thought they could use it to turn lead into gold

1 In 1829, Robert Brown was using a microscope to look at some dead pollen grains floating in water. The pollen grains were jerking about in all directions.

2 Brown couldn't explain what was happening. Much later, in 1909, Albert Einstein explained that the pollen grains were moving because they were being bombarded by water molecules.

3 This **Brownian motion** gave good evidence of moving water molecules. Later it was used to do the first calculation of the size of a molecule.

1 What is the difference between a guess and a scientific theory?
2 a) Why do pollen grains move when they are floating in water? ▲
   b) Think up experiments to demonstrate that the pollen grains are not being moved by draughts, or being attracted to the glass.
3 When you look at smoke particles under a microscope, they are moving about. What causes this?
4 Iron atoms do not have hooks, but can you suggest why the Greeks thought that they had?

Did you know?

- Alchemists invented lots of recipes for making gold. One recipe included mercury, coal and the yolks of 50 eggs! The whole lot was then buried in horse dung! It didn't work...

4.3 Air pressure

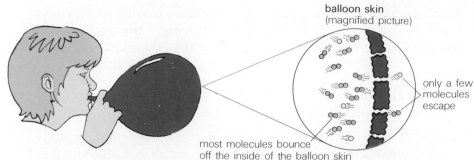

balloon skin
(magnified picture)

only a few
molecules
escape

most molecules bounce
off the inside of the balloon skin

When you blow up a balloon, you blow millions and millions of molecules into it. These molecules fly around inside the balloon in all directions. Many of them bounce off the inside of the balloon rubber.

Air pressure

Whenever a molecule bounces off the balloon rubber, it gives the rubber a tiny push. Millions of tiny pushes add up to give the **air pressure** which stretches the balloon skin.

Air pressure is produced by air molecules bouncing off a surface.

1. Oil can (oil emptied out) The can is full of air. This air presses out. The air outside presses in. The can keeps its shape because the air pressure on the inside is equal to the air pressure on the outside

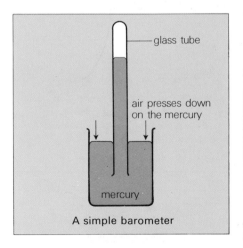

air pumped out

2. Oil can (really empty!) There is almost no air inside to press out. The air pressure on the outside makes the can collapse

There are air molecules bouncing off you, and the things around you. Everything is affected by air pressure. The bigger the number of air molecules bouncing off the surface, the bigger is the air pressure on it.

It's amazing how much pressure these tiny molecules can produce! You can make a strong can collapse by pumping air out of it, as shown in the illustrations above.

Barometers are used to measure air pressure. You can make a simple barometer from a piece of glass tubing (sealed at one end), and a dish of mercury. First, you fill the tube with mercury. Then you quickly turn it upside down so that the open end is under the mercury in the dish. Air pressure prevents most of the mercury from leaving the tube. The height of the column that remains depends on the air pressure.

glass tube

air presses down
on the mercury

mercury

A simple barometer

1. What produces air pressure? When is the air pressure on a surface large? ▲
2. a) Why does the used oil can keep its shape?
   b) Why does the can collapse when the air is pumped out? ▲
3. Do you think that the air pressure is greater inside or outside a blown up balloon? Suggest a reason for your answer.
4. What is a barometer used for? ▲
5. Explain what will happen to the height of the mercury column in the simple barometer: a) when air pressure rises b) when air pressure falls.
6. **Try to find out:** a) why airliners have pressurised cabins
   b) what an aneroid barometer is, and how it works.

Did you know?

- Changes in air pressure affect the weather.
- The air pressure pressing on you is roughly the same as the pressure on an arm chair when an elephant sits on it.

4.3 Expansion

Solids, liquids and gases all take up more space or **expand** when you heat them. Heating makes their atoms and molecules move faster. And this makes the atoms and molecules move further apart.

1 Gases expand:

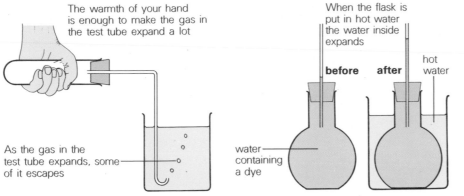

The warmth of your hand is enough to make the gas in the test tube expand a lot

As the gas in the test tube expands, some of it escapes

2 Liquids expand:

When the flask is put in hot water the water inside expands

before   after   hot water

water containing a dye

3 Solids expand:

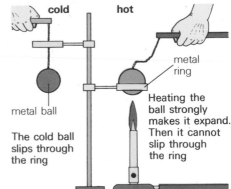

cold   hot

metal ring

metal ball

The cold ball slips through the ring

Heating the ball strongly makes it expand. Then it cannot slip through the ring

If you cool a solid, a liquid or a gas, the opposite happens. They all get smaller, or **contract**.

Effects of expansion and contraction

Many substances produce very large forces when they expand and contract. Expansion in very hot weather has bent railway lines and cracked concrete pavements. Contraction in very cold weather has snapped telephone wires. But it is usually possible to prevent damage like this from happening.

Different substances expand by different amounts. Gases expand most, then liquids, then solids.

The expansion of solids is very small. The two metals in the **bimetallic strip** on the right expand by less than 1 mm when they are heated in a bunsen flame. But they expand by different amounts, and that makes the strip bend.

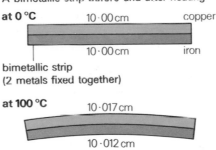

A bimetallic strip before and after heating

at 0 °C   10·00 cm   copper
              10·00 cm   iron

bimetallic strip
(2 metals fixed together)

at 100 °C   10·017 cm

              10·012 cm

The strip bends because copper expands more than iron

The **expansion joints** in bridges are hard to construct. The ones in this road bridge are faulty and will have to be replaced

1 'Water expands when it is heated.' What does this mean? ▲
2 Why should large concrete slabs be laid with expansion joints so that they are not touching? ▲
3 What is a bimetallic strip? Why does it bend when heated? ▲
4 Engineers who fix up new telephone wires in summer leave the wires hanging slack. Why?
5 Why must oven shelves not fit tightly when the oven is cold?
6 **Try to find out:** a) how a thermometer works   b) how metal rims are fitted to trains' wheels   c) how a bimetallic strip and a bell can be used to make a fire alarm.

Did you know?

- Gases expand roughly 3000 times more than solids.
- The Railway Bridge over the River Forth is made of iron. It is more than ½ metre longer in summer than in winter.

41

4.3 Density

Going further

Floating and sinking can be puzzling. Why does a light copper coin sink when it is put in water? Why does a much heavier oak log float?

Why does this happen ?

You will realise straight away that it can't be the mass of the object which makes it float or sink. In fact it is the object's **density** which matters. Copper has a greater density than water, and so the coin sinks. Oak wood has a smaller density than water and so the log floats. Density can be defined in this way:

The density of a substance is the mass of 1 cm³ of it.

The unit of density is **grams per cubic centimetre**, written **g cm⁻³**. Iron has a density of $7.8\,\mathrm{g\,cm^{-3}}$. This means that each cubic centimetre of iron has a mass of $7.8\,\mathrm{g}$.
You can find the density of any object if you know its mass and its volume.

$$\text{density} = \frac{\textbf{mass}}{\textbf{volume}} \frac{\text{(in grams)}}{\text{(in cm}^3)}$$

The oak wood log has a volume of $2000\,\mathrm{cm^3}$ and mass of $1800\,\mathrm{g}$ and so

density of oakwood $= \dfrac{1800\,\mathrm{g}}{2000\,\mathrm{cm^3}} = 0.9\,\mathrm{g\,cm^{-3}}$.

The density of a substance depends on two things:

- the mass of its atoms (or molecules)
- how closely these atoms (or molecules) are packed together.

Gold is made up of heavy atoms closely packed together and so it has a high density. On the other hand, gases have very low densities. Gas molecules spread out to occupy a large volume, with lots of empty space. $1\,\mathrm{cm^3}$ of *any* gas is very light.

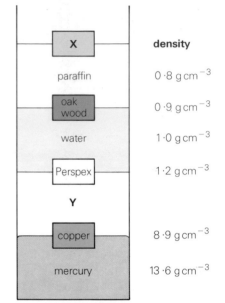

	density
X	
paraffin	$0.8\,\mathrm{g\,cm^{-3}}$
oak wood	$0.9\,\mathrm{g\,cm^{-3}}$
water	$1.0\,\mathrm{g\,cm^{-3}}$
Perspex	$1.2\,\mathrm{g\,cm^{-3}}$
Y	
copper	$8.9\,\mathrm{g\,cm^{-3}}$
mercury	$13.6\,\mathrm{g\,cm^{-3}}$

A density ladder

Did you know?

- Sea water is denser than fresh water. That's why it's easier to float in the sea than in a lake.
- The Dead Sea is so dense that you can float sitting up in it.

1. What is the density of a substance? ▲
2. Water has a density of $1\,\mathrm{g\,cm^{-3}}$. What does this mean? ▲
3. Why does the copper coin sink when put in water? ▲
4. Look at the density ladder:
    a) Why does the oak wood float on water but not on paraffin?
    b) Suggest densities for **X** and **Y**, explaining each choice.
    c) The density of nylon is $1.1\,\mathrm{g\,cm^{-3}}$. Where on the ladder would a piece of nylon float? Explain your answer.
5. What is the density of: a) a brass nut, mass $34\,\mathrm{g}$, volume $4\,\mathrm{cm^3}$ b) a cork, mass $2\,\mathrm{g}$, volume $8\,\mathrm{cm^3}$?
6. Why do bubbles of gas rise through lemonade?
7. **Try to find out:** exactly how a hot air balloon works.

Hot air has different density to cold air. That's why hot air balloons rise

For the enthusiast

Here are some problems for you to work out. They are problems about expansion, pressure and density. You may have to hunt in other books to answer some of the questions, particularly the ones marked *

1 Blowing up a beach ball
When you are blowing up a beach ball:
a) what are you blowing into the ball?
b) what happens to the ball?
c) why does this happen?

2 Heating a sealed tin
a) What will happen to the tin lid?
b) Why will this happen?
c) What warning is written on an aerosol can?*
d) Why is a warning needed?

3 Drinking lemonade
a) What happens to the air pressure in the straw when you suck?
b) Why does this happen?
c) What pushes the lemonade up the straw?

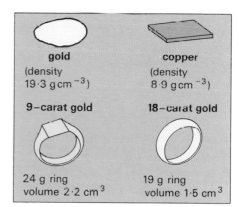

4 A simple thermostat
Switching on the electricity makes the heater glow. A few minutes later, a gap opens up at **A**. The electricity and the heater go off.
a) Why does a gap open up?
b) Why will the heater keep going on and off?
The bimetallic strip makes a simple **thermostat**.
c) What does a thermostat do?*
d) Where are thermostats used?*

5 Releasing a weather balloon
This photograph shows a weather balloon carrying instruments for measuring the weather. In the upper atmosphere, air pressure is much lower than at sea level.
a) How will the weather balloon's volume change as the air pressure round it gets less? Explain.
b) 16 km up, the balloon will burst. Why?
c) Why does breathing get more difficult at higher altitude?*

6 Checking the purity of gold
Copper is mixed with gold to make a harder metal.
a) Work out the density of the metal in each ring.
b) Which ring has more copper in it? Explain.
c) A 24 carat gold ring will have a density of $19.3\,\mathrm{g\,cm^{-3}}$. What does the number of carats tell you about a piece of gold?*

4.4 Elements

If you heat a piece of bread, or a piece of wood, or even a piece of fingernail for long enough, you will be left with a black solid. That solid is called carbon. It is one of the simplest chemicals. It is called an **element**. The word 'elementary' means 'simple'.

An element is a chemical substance which can't be broken down into anything simpler. Carbon is called an element because it is only made up of carbon atoms.

Copper is another example of an element. It is made up only of copper atoms. There are just over 100 different elements. Each has its own kind of atom. Around 90 elements have been found in nature. The rest have been made by scientists.

Each element has a **symbol**. 'H' is the symbol for hydrogen. 'He' is the symbol for helium. You will find a list of all the elements, with their symbols, on a chart called the **Periodic Table of the Elements**. You can see a Periodic Table, and photographs of a few of the elements, below. Some elements can't be seen in a photograph. Oxygen, hydrogen and nitrogen are three of the elements which are invisible gases.

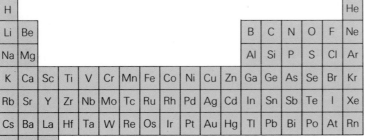

The Periodic Table

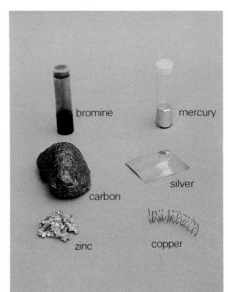

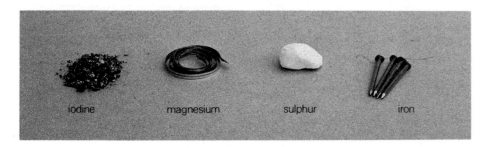

1 What is: a) an element b) the Periodic Table? ▲
2 Why is carbon an element? ▲
3 How many kinds of atom are in a lump of copper? Explain. ▲
4 Most elements are solids, a few are gases, two are liquids. Divide the elements *mentioned on this page* into 3 sets, solids, liquids and gases.
5 Carbon is a black solid.
   Describe: a) sulphur b) mercury c) copper.
6 Which of the elements in the photographs is used: a) in jewellery b) in thermometers c) in plumbing?
7 **Try to find out:** what some of the other elements are used for.

Did you know?

- Diamond and charcoal are two different forms of the element carbon.
- The biggest diamond ever found weighed almost 600 g. That meant that it contained about 30 million, million, million, million atoms. But it only contained carbon atoms.

4.4 Compounds

Iron and sulphur are two of the elements.

1 You can mix powdered iron and powdered sulphur without making a new chemical. You can still see the sulphur in the mixture. You can still pull the iron away with a magnet.

2 But if you heat a mixture of iron and sulphur, a glow passes through the mixture. This time a new chemical is formed.

3 The new chemical is called **iron sulphide**. It contains sulphur (but you can't see any yellow in it). It contains iron (but you can't pull it away using a magnet).

Chemical reactions and compounds

When iron and sulphur are mixed, the atoms don't join up. But when the mixture is heated, a **chemical reaction** takes place. The iron and sulphur atoms do join up. That's when the new chemical is formed.

Iron sulphide is called a **compound** because it contains iron atoms joined to sulphur atoms.

Any chemical which contains two or more elements joined together is called a compound.

Most chemicals are compounds. The photographs shown here illustrate three common compounds. A compound's name may tell you which elements are joined up in it. **Copper iodide**, for example, contains **copper** and **iodine**.

Sand   (silicon dioxide)

Common Salt   (sodium chloride)

Water   (hydrogen oxide)

1 What is a compound? ▲
2 Heating iron and sulphur produces a chemical reaction. What happens in the reaction? ▲
3 Why is it easier to separate atoms from a mixture than from a compound?
4 Which elements are joined up in: a) sand b) salt c) water?
5 **Try to find out:** the name of the main compounds found in:
    a) ruby b) rust c) the 'rotten egg' gas.

Did you know?

- Over one million chemical compounds are known.
- Some compounds contain five or more different elements joined together.

45

4.4 Joining up makes a difference

Hydrogen is an element. It is a gas which burns well. It can even explode when you light it.

Oxygen is another element. It is a gas which helps things to burn.

Hydrogen and oxygen join to give the compound **hydrogen oxide**. You know it better as water. It puts out fires.

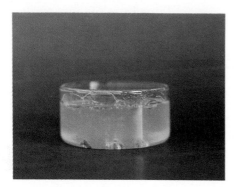

The element **calcium** bubbles and fizzes in water.

Phosphorus is a rather nasty element. It catches fire in air when it gets dry.

Calcium, phosphorus and oxygen are joined in the compound **calcium phosphate**, found in teeth. Luckily it does not fizz in water or burn when dry!

As you can see, compounds are usually very different from the elements which make them up. And that's just as well. Your body is made up of a number of elements, some of which are very dangerous. These dangerous elements include chlorine (a poisonous gas), iodine (a poisonous solid), sodium and potassium (which fizz in water) as well as calcium, hydrogen and phosphorus. Fortunately, your body contains these elements joined up in harmless compounds!

1 a) What is the chemical name for water? ▲
   b) How is water different from the elements which make it up? ▲
2 a) Which chemical compound is found in teeth? ▲
   b) What would happen to your teeth if they were made of the element calcium? Why?
   c) Milk is good for making teeth because it contains calcium. Do you think that milk contains calcium as an element or a compound? Explain your answer.
3 **Try to find out:** what phosphorus element is used for.

Did you know?

- Bones contain calcium phosphate.
- A compound can be more dangerous than the elements which make it up. Hydrogen sulphide – rotten egg gas—is much more poisonous than either hydrogen or sulphur.

46

When a piece of very thin copper foil is put into chlorine gas, there is a flash of flame. A chemical reaction takes place. A blue green solid called **copper chloride** is produced. Heat and light energy are produced, too. You can write down what happens in the reaction like this:

copper + chlorine ⟶ copper chloride + energy (heat and light)

This 'shorthand' way of writing down what happens is called a **chemical equation**.

The heat and light are produced when the copper and chlorine atoms join together. If you want to split up the copper chloride to get the copper and chlorine atoms back again, you have to supply energy. Electrical energy will do this:

copper chloride + electrical energy ⟶ copper + chlorine

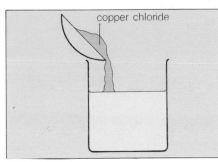

1 The copper chloride is dissolved.

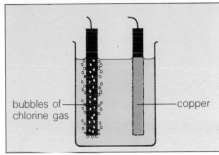

2 Carbon rods are put in the solution and connected up to an electrical supply.

3 As the current passes, solid copper and chlorine gas are produced.

In many reactions between elements, energy is produced.
In these reactions, energy may have to be supplied to start the reaction off. But once it has started, a lot more energy is produced.

To split up compounds, energy is needed. The energy is needed to separate the joined atoms.

Did you know?

- Aluminium, calcium, sodium and magnesium are all produced from their compounds by using electricity.
- A reaction in which electricity is used to break up a compound is called an **electrolysis**.

1 How do you know that a chemical reaction takes place when thin copper foil is put in chlorine? (Give 2 pieces of evidence.) ▲
2 Why is energy needed to split up a compound? ▲
3 What is meant by an electrolysis? ▲
4 a) Describe what you would do to produce copper and chlorine from copper chloride. ▲
   b) Is this reaction an electrolysis? Explain your answer.
5 **Try to find out:** how silver plate is put on teaspoons.

A huge amount of electricity is used to release aluminium from its compounds. This factory is in Sweden, where electricity is cheap

Solvents and solutions

The sea is the world's biggest *solution*. It covers 70% of the Earth's surface.

Water is the *solvent* in the sea, but you can't use sea water for drinking. (If you have ever swallowed a mouthful of sea water while swimming, you will be able to guess why!) The sea has very many different chemical *solutes dissolved* in it.

Every day, huge quantities of water *evaporate* from the sea. Despite this, the sea's level does not go down. The *water vapour condenses* and forms rain. Then the water runs back into the sea. That's what happens in the *water cycle*.

Did you understand everything in the passage you have just read?

Don't worry if the answer is 'no'. This section's job is to teach you about solvents, solutes, solutions, evaporation, condensation, dissolving and about other things, too.

It might be a good idea to read this page again when you have finished the section. You should understand it better then!

5.1 Water and the water cycle

Water is usually found as a liquid, but it can also be found as a solid and a gas. The solid is called **ice**. **Steam** and **water vapour** are names used for the gas.

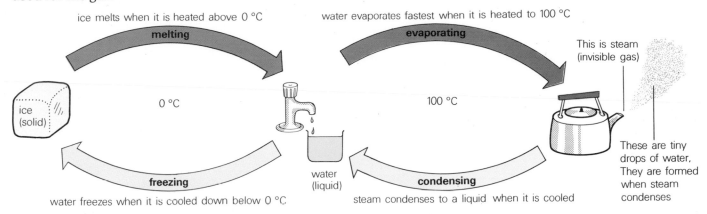

ice melts when it is heated above 0 °C

melting

water evaporates fastest when it is heated to 100 °C

evaporating

This is steam (invisible gas)

ice (solid)

0 °C

100 °C

These are tiny drops of water, They are formed when steam condenses

freezing

water freezes when it is cooled down below 0 °C

water (liquid)

condensing

steam condenses to a liquid when it is cooled

0 °C is called the **melting point** of ice. It is also the **freezing point** of water! 100 °C is the **boiling point** of water. Water evaporates quickly when it is boiling at 100 °C. It can also evaporate at lower temperatures, but then evaporation is slower.

The water cycle

Melting, freezing, evaporating and condensing all play a part in the **water cycle**. (A cycle goes round and round – so does the world's water!)

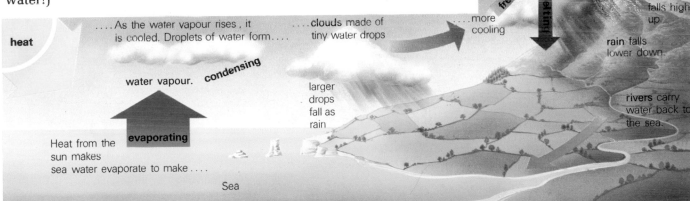

clouds of ice crystals

freezing

melting

snow falls high up.

heat

....As the water vapour rises, it is cooled. Droplets of water form....

...clouds made of tiny water drops

...more cooling

rain falls lower down.

water vapour. **condensing**

larger drops fall as rain

rivers carry water back to the sea.

Heat from the sun makes sea water evaporate to make....

evaporating

Sea

1 What is produced when: a) ice melts b) water evaporates c) water freezes d) steam condenses? ▲
2 In melting, a solid changes to a liquid. What happens in: a) freezing b) evaporating c) condensing? ▲
3 The world's water goes round and round in the water cycle. Explain what happens.
4 What is each of the following? Where would you find one?: a) a large block of floating ice b) a jet of hot water coming from the ground c) a mid-air collection of water droplets d) a large mass of ice moving down a mountain?
5 **Try to find out:** how much rain falls on your town in a year.

Did you know?

- ¾ of the world's rain falls over the sea.
- 98% of the world's water is in the form of liquid. About 2% is in the form of ice. Only 0.0004% is in the form of water vapour in the atmosphere.

5.1 Icy winters in Britain

Frost and snow change the surroundings in lots of different ways. This is because water changes to ice when the temperature falls to 0 °C. Here are four facts which will help you to understand what happens in a freeze-up:

1 Water *expands* when it freezes.

2 This expansion means that ice is *less dense* than water.

3 Salty water does not freeze until the temperature is below 0 °C.

4 Ice melts when pressure is put on it.

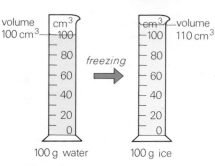

volume 100 cm³ — 100 cm³ / *freezing* → cm³ — volume 110 cm³

100 g water   100 g ice

Water expands when it freezes

Skaters really glide on water (see question 3)

Water needs space to expand when it freezes. Thats why pipes burst.

Ice floats on top of a pond because it is less dense than water

Salt is put on roads in winter. This makes sure that ice will not form till the temperature falls well below 0 °C

When you make a snowball, you press on the snow. Some of the snow melts to water. When you stop pressing, the water freezes again. The new ice keeps the snowball together

1 When water changes to ice, what happens to: a) its volume b) its density? ▲
2 Why does: a) freezing water burst pipes b) ice float on top of a pond c) pond water freeze more easily than sea water?
3 Explain why salt is put on roads in winter. ▲
4 When a skater passes over a piece of ice, the ice melts. Then the water freezes again. Why does this happen?
5 Burst pipes usually go unnoticed till a thaw comes. Explain why.
6 **Try to find out:** how to protect your home against frost.

Did you know?

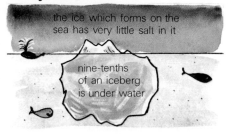

the ice which forms on the sea has very little salt in it

nine-tenths of an iceberg is under water

5.1 Typical British weather

Cloudy ...

The cloudiest, wettest parts of Britain are round the mountains on the west coast. Large amounts of cloud and rain are produced when warm, moist air from the Atlantic Ocean is blown up and over the barrier of high hills.

Low lying areas can get very wet, too. Heavy rain is often produced at a **front** where warm air and cold air meet.

If fast moving warm air overtakes slower moving cold air, the warm air rises. High wispy **cirrus** cloud is produced first. This is followed by lower, thicker **stratus** cloud which often produces hours of steady rain.

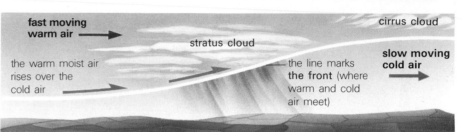

fast moving warm air →

stratus cloud

cirrus cloud

the warm moist air rises over the cold air

the line marks **the front** (where warm and cold air meet)

slow moving cold air →

... with rain, snow and hail

In both of the cases described above, the clouds and rain are produced when the warm, moist air rises. Warm air can hold far more water vapour than cold air. When warm, moist air rises and cools down, some of the water vapour changes to tiny drops of water or crystals of ice.

1 The water drops and ice crystals first formed are so light that they cannot fall. They float in the air, forming clouds.

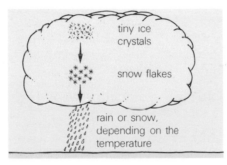

tiny ice crystals

snow flakes

rain or snow, depending on the temperature

2 But ice crystals can join together to make a snowflake heavy enough to fall. This is how most **precipitation** (rain, snow or hail) is produced.

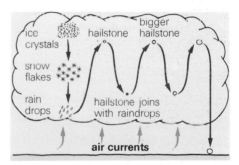

ice crystals

snow flakes

rain drops

hailstone

bigger hailstone

hailstone joins with raindrops

air currents

3 Hailstones are produced in clouds with strong air currents. They are really frozen rain drops which are tossed up and down several times before they fall.

1 When does warm, moist air produce clouds? ▲
2 Why is rain produced: a) on Britain's west coast b) at a front? ▲
3 What is meant by precipitation? How is it produced? ▲
4 Snowflakes will not fall if the air temperature is 5 °C but hailstones often do. Why?
5 Would you expect cirrus clouds to be made of water drops or ice crystals? Explain.
6 **Try to find out:** a) what is meant by cloud seeding b) about other types of cloud.

Did you know?

- The temperature at the top of a cloud can be as low as −40 °C.
- A hailstone gets an extra coat of ice each time it is tossed upwards. The biggest hailstone ever measured was 19 cm across.

5.2 More about evaporation

When water **evaporates**, it changes to a gas. This can happen at any temperature between 0 °C and 100 °C. But evaporation does not always take place at the same rate. Sometimes it is fast and sometimes it is slow.

If you want to dry wet things quickly, you can:

1 heat them	**2 spread them out to dry**	**3 blow air over them**

Heat energy is needed to change water from a liquid to a gas. That's why water evaporates more quickly when it is heated.

Water evaporates more quickly from a large surface than from a small one. That's why wet things dry faster when they are spread out.

Moving air carries away water vapour as soon as it is formed. That's why things dry faster when air is blowing over them.

High temperatures, large surfaces, and moving air all help with drying. They all speed up evaporation.

Dried milk is produced from ordinary milk by evaporating off the water. Evaporation has to be quick. The milk is spoiled if it is heated for too long.

The dried milk is produced in a **spray drier**. The milk is sprayed in as tiny drops. These tiny drops have a very large surface area. When they meet the hot air flowing through the drier, the water evaporates from them immediately. This forms a powder – dried milk – which falls to the bottom of the drier.

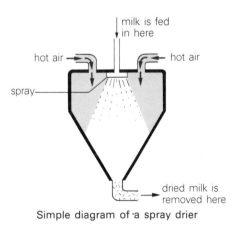

Simple diagram of a spray drier

1 What happens to water when it evaporates? ▲
2 Give three ways in which evaporation can be speeded up. ▲
3 Which kind of weather helps puddles to dry? Why?
4 a) How is evaporation speeded up in a spray drier?
  b) Why must the water in the milk be evaporated quickly?
5 How does a hair drier help to dry hair quickly?
6 **Try to find out:** how other dried foods are produced.

Did you know?

● Milk is 87% water.
● Dried food keeps longer. The bacteria which spoil food can't grow without water.

In the world's **arid lands**, the soil is very dry — too dry for growing crops. There is not much rain. Even when rain *does* fall, the hot, dry, windy weather quickly evaporates any moisture in the soil.

The water shortages in many of these areas have been overcome by bringing in extra water. This is called **irrigation**. But water loss by evaporation is still a big problem. Here are four ways in which evaporation is cut down and water is saved:

film of cetyl alcohol covers the water in the reservoir

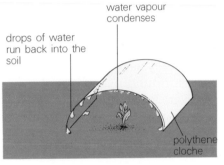

water vapour condenses

drops of water run back into the soil

polythene cloche

1 When water is stored in reservoirs, the surface can be covered with a film of waxy solid called **cetyl alcohol**. This prevents the water underneath from evaporating.

2 Plants are grown in cloches to save water. The sun's heat does make water evaporate from the soil. But the water can't escape. It condenses on the walls of the cloche and runs back into the ground.

3 If the irrigation water flows through open ditches, much of it evaporates. That's why the water is carried through pipes wherever possible.

4 Piling stones round the foot of each plant is a simple way of preventing the soil from drying up. A pile of stones may even help to add a little water. At night, the stones cool down very quickly. Water vapour from the air condenses on their cold surfaces. This produces tiny drops of dew which trickle down into the soil.

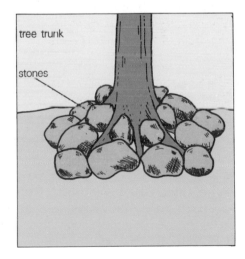

tree trunk

stones

Did you know?

- About 400 000 000 000 000 000 litres of water evaporate from the Earth each year.
- Plans are being made to tow icebergs from the Antarctic to Australia and the Middle East to irrigate the land.

1 What is meant by irrigation? Why is it important? ▲
2 Explain how water can be saved by using:
   a) cetyl alcohol b) cloches c) pipes for carrying irrigation water. ▲
3 How does dew form?
4 Would a wide, shallow reservoir or a narrow, deep one be better for storing water in a hot country? Explain.
5 The emergency water supply shown in the diagram opposite could save your life in a desert. How does it work?
6 **Try to find out:** a) the names of some of the world's arid lands
   b) what mulches are, and why gardeners use them.

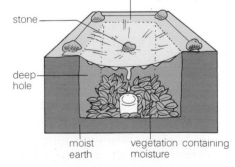

sheet of polythene fixed in place by stones

stone

deep hole

moist earth

vegetation containing moisture

5.2 Cooling by evaporation

A drop of water is a mass of tiny, moving water molecules. The molecules in the drop all have different energies. They are all moving at different speeds. Some low energy molecules move very slowly. Some high energy molecules move very rapidly. Most molecules have a speed somewhere in between.

The drop is held together because the water molecules pull on or **attract** each other. But this attraction is not strong enough to keep every molecule in the drop. Some high energy molecules near the surface move so fast that they escape. As this happens, the drop slowly evaporates.

Only the fastest moving molecules have enough energy to escape from the drop. But evaporation does not stop once they have gone. Instead, the water drop takes in heat energy from its surroundings (if it can). This energy speeds up the molecules which are left in the liquid, and more escape. All of this means that:

when a liquid evaporates, it takes in heat from its surroundings and makes the surroundings colder.

This cooling effect is used in a refrigerator. The refrigerator's pipes contain **freon**, a substance which evaporates and condenses easily.

In a liquid

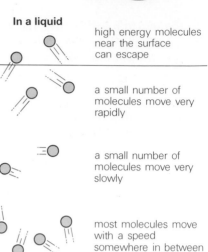

high energy molecules near the surface can escape

a small number of molecules move very rapidly

a small number of molecules move very slowly

most molecules move with a speed somewhere in between

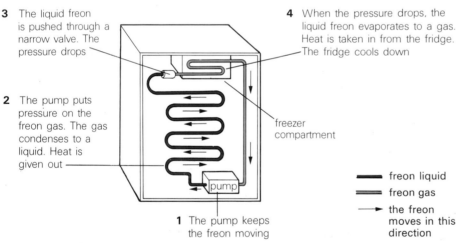

3 The liquid freon is pushed through a narrow valve. The pressure drops

4 When the pressure drops, the liquid freon evaporates to a gas. Heat is taken in from the fridge. The fridge cools down

2 The pump puts pressure on the freon gas. The gas condenses to a liquid. Heat is given out

freezer compartment

1 The pump keeps the freon moving

pump

— freon liquid
= freon gas
→ the freon moves in this direction

The working parts of a refrigerator

Rear view of fridge. You can see the pump and the tubes through which warm freon flows

1 Complete these sentences:
   a) A water drop is held together because . . . .
   b) A water drop slowly evaporates because . . . .
   c) When a water drop evaporates, the surroundings become . . . because . . . . ▲
2 Is energy given out or taken in when: a) a gas changes to a liquid b) a liquid changes to a gas? ▲
3 Where does the cooling take place in a refrigerator? Why does cooling take place there?
4 Why does a refrigerator warm the room it is in?
5 Why do you feel cold when you stand around wet after swimming?
6 **Try to find out:** about some old fashioned methods of keeping food cool.

Did you know?

- Heat is given out when a gas condenses.
- One large store uses the heat given out by its freezers to heat the water needed by its hairdressing salon.

When you put a crystal of sugar into water it **dissolves**. The water separates the sugar molecules which were joined up in the crystal. The sugar molecules spread through the water.

Because it dissolves, sugar is described as **soluble**. Together, sugar and water make a **solution**.

moving water molecules

molecules in grain of sugar

water (solvent)

magnified picture

grain of sugar (solute)

sugar molecules and water molecules mixed together

sugar solution

magnified picture

● = sugar molecule
○ = water molecule

A solution is made whenever two substances mix completely through each other. The substance which dissolves is called the **solute**. The substance which does the dissolving is called the **solvent**. (In this solution, sugar is the solute. Water is the solvent.)

Sugar solutions are used as the 'syrups' in tins of fruit. Here is a recipe for making a syrup using water and granulated sugar (sugar made up of small crystals):

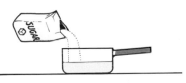

1 Add 500 grams of sugar to 1 litre of water in a saucepan.

2 Heat the mixture, increasing the temperature slowly . . .

3 . . . stirring all the time.

A busy housewife does not have time to wait for the sugar to dissolve in cold water. She heats the mixture because solutes dissolve faster in hot water. Stirring speeds up dissolving, too. She *could* use sugar lumps instead of granulated sugar but that would take longer: small crystals dissolve faster than large lumps.

Did you know?

- A substance which does not dissolve is **insoluble**. But there is no substance which is *completely* insoluble.
- Sugar solutions prevent tinned food from going bad.

1 What is meant by: a) a solute b) a solvent c) a solution d) an insoluble substance e) saying that sugar is soluble? ▲
2 What happens when a sugar crystal dissolves?
3 Name: a) the solvent b) one solute in the sea.
4 Give three ways in which dissolving can be speeded up. ▲
5 In the experiment opposite, the crystals were added at 9.00 a.m. exactly. The crystals in one beaker dissolved completely by 9.05 a.m. The crystals in the other beakers dissolved by 9.15, 10.00 and 11.00 a.m. Match up the beakers and times, explaining each choice.
6 **Try to find out:** the solutes in Coca-Cola (try looking at a big bottle!).

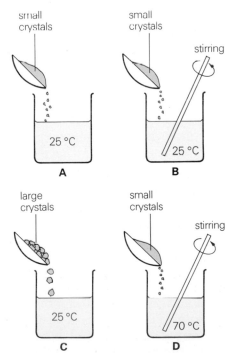

small crystals

small crystals

stirring

25 °C

25 °C

A

B

large crystals

small crystals

stirring

25 °C

70 °C

C

D

All the beakers contain 100 cm³ water.
5 g copper sulphate is added to each

5.3 Crystals from solutions

Blue copper sulphate crystals dissolve well in water. You can dissolve 24 g of crystals in 100 cm³ of water at room temperature. After 24 g has been added, however, no more solid will dissolve. You have made a **saturated** solution – one which will dissolve no more solute.
You can learn more about saturated solutions from the experiment shown below:

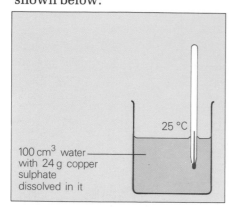

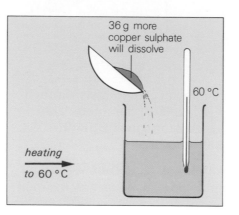

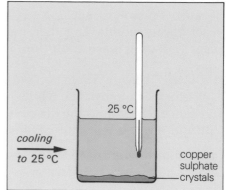

Cold saturated solution
At 25 °C, **24 g** of copper sulphate will dissolve.

Hot saturated solution
At 60 °C, **60 g** of copper sulphate will dissolve.

Cold saturated solution
Cooling to 25 °C, **24 g** stay dissolved. **36 g** form crystals.

The experiment shows that:

1 hot water can dissolve more solid than cold water. By heating the water from 25 °C to 60 °C, it is possible to dissolve 36 g more copper sulphate.

2 crystals form when a hot saturated solution cools down. The solution at 60 °C dissolves 60 g copper sulphate. By the time it has cooled down to 25 °C, only 24 g are still dissolved. The other 36 g form crystals at the bottom.

Slow cooling produces the biggest crystals. In the photograph, you can see crystals of copper sulphate made by rapid cooling (1) and by slow cooling (2).

The photograph also shows crystals of other chemicals. Crystals of different chemicals may have different shapes, but all have smooth faces and sharp edges.

1 What is a saturated solution? How would you make one? ▲
2 How could you make crystals of copper sulphate from copper sulphate powder? How could you make sure that the crystals were large?
3 Finish the sentence: 'All crystals have ...'. ▲
4 Can you make a saturated solution of copper sulphate by adding: a) 13 g of crystals to 50 cm³ of water b) 500 g of crystals to 1 litre of water? (The water is at 25 °C.)
5 Copper sulphate crystals grow from saturated solutions, but they are not thought to be alive. Why not? (look back to page 13)
6 **Try to find out:** the names of some gemstone crystals.

Did you know?

- Granulated sugar is produced by cooling down hot sugar solutions quickly. Rapid cooling gives small crystals.
- Scientists use X-rays to find out more about crystals.

56

5.3 Colloids

When sugar and water are mixed, a **solution** is produced. The solution is clear. The sugar is spread evenly through the water.

When powdered chalk and water are mixed, a **suspension** is produced. This suspension is milky. The chalk is only spread evenly through the water while mixing takes place. When mixing stops, the chalk settles out at the bottom.

When powdered starch is mixed with water, wallpaper paste is produced! The paste is not a true solution (although like the solute in a solution, the starch stays spread through the water and does not settle out). The paste is not a suspension either. In fact, wallpaper paste is one example of a **colloid**.

A colloid is made up of two substances which don't dissolve in each other. It consists of tiny particles of one substance spread through the other. The particles spread through a colloid are all between 0.001 cm and 0.000 000 1 cm in diameter. At that size, they are too large to dissolve but too small to settle out.

As you can see from the examples opposite, solids, liquids and gases can all be involved in making colloids. **Emulsions** are colloids which consist of tiny droplets of one liquid spread through another. A **foam** consists of tiny bubbles of gas spread through a liquid.

Milk: tiny droplets of liquid fat spread through water

Emulsion paint: tiny drops of oily paint spread through water

Ice cream: tiny bubbles of air spread through 'milk ice'

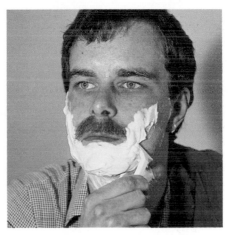

Shaving cream: tiny bubbles of air spread through soap

Some everyday colloids

1 What is: a) a colloid b) an emulsion c) a foam? ▲
2 Pick out one emulsion and one foam from the pictures above.
3 In what way is a colloid: a) the same as b) different from a solution? ▲
4 The particles and droplets in a colloid never settle. In the case of emulsion paint, this is useful. In the case of a room full of smokers, this is a nuisance. Explain the difference.
5 Many indigestion mixtures are suspensions. What do the instructions tell you to do before taking the mixture, and why?
6 **Try to find out:** how ice cream is made.

Did you know?

- You can't separate solid particles from a colloid by filtering it. The particles are so small that they pass through the filter.
- 50% of the volume of ice cream is air.

5.3 More or less soluble

The sea contains different chemicals dissolved in its water. When sea water evaporates, these chemicals can form rocks called **evaporites**.

Evaporites form in hot regions of the world such as the edges of the Red Sea. They can be formed in different ways. This is one of the simplest:

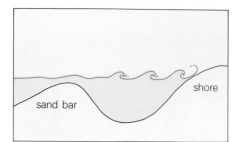

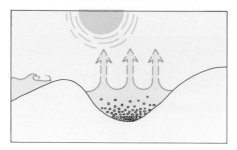

1 As the waves come in to the shore, they build up a sand bar. Sea water is trapped between the sand bar and the shore.

2 The trapped water slowly evaporates in the hot Sun. The water becomes saturated. When water evaporates, solid forms and falls to the bottom.

3 The water evaporates completely, leaving layers of minerals – calcite, gypsum and salt. These minerals make up the evaporite.

In evaporites formed in this way, the layers of minerals are always found in the same order (calcite at the bottom, then gypsum, then salt). This happens because calcite is less soluble than gypsum and gypsum is less soluble than salt. When the water evaporates, the least soluble substance (the calcite) forms solid first.

Is anything really insoluble?

You probably won't be surprised to learn that some substances are more soluble than others. But you may be surprised to learn that calcite is the same chemical as chalk. No doubt you thought that chalk was completely insoluble in water.

In fact, most substances dissolve in water, although some substances only dissolve by a tiny amount. Many substances which are normally thought to be completely insoluble are really very slightly soluble. Silica, the main chemical in sand, is one of these. Only 0.01 g of silica dissolves in 100 cm³ cold water, but much more silica will dissolve in hot water under pressure. If a hot, high pressure silica solution cools down slowly, large crystals can be produced. These are crystals of **quartz**, another form of silica. Some of the quartz crystals found in rocks have been formed in this way.

These large crystals of quartz were formed naturally

1 What is an evaporite? How is an evaporite formed? ▲
2 Why is calcite always found at the bottom of an evaporite? ▲
3 Is sand really insoluble? Suggest why it is said to be insoluble.
4 Water which flows through volcanic ground has more silica in it than any other water. Suggest why.
5 Explain how you could separate: a) salt and sand b) salt and sugar. (Clue: salt is more soluble than sugar)
6 **Try to find out:** the names of other compounds which are dissolved in sea water.

Did you know?

- If the sea was completely evaporated, there would be enough solid to cover the Earth's surface to a depth of 45 m.
- Sponges take silica from sea water and use it to build up their bodies.

5.4 Using different solvents

Water is by far the most important solvent. It dissolves very many solutes. It is cheap and safe, and there is plenty of it!

Water is used for lots of different jobs around the home, but more water is used for washing than for anything else. Water is a very useful solvent for washing. By itself it can dissolve many stains. It can dissolve soaps and detergents, and with them is even better at cleaning things.

Different solvents for different jobs

But even with detergent or soap, water can't be used to clean everything. There are many substances which won't dissolve. Water can't be used to dissolve dirty oil from a car engine, or chewing gum or nail varnish. Other solvents have to be used to dissolve these substances. **Paraffin** dissolves oil, so mechanics can use it to clean car engines. **Acetone** dissolves nail varnish so it can be used in nail varnish remover. **Trichloroethane** dissolves chewing gum and is used to remove it from clothes.

If you want to remove a stain from clothes, you should first try to wash it out with soapy water. If that does not work, you could try using other solvents. (**Meths** and **white spirit** can be useful here.) But you must know what you are doing! Some solvents can dissolve man-made fibres!

It's often safer to take the clothes to the **dry cleaners**. Dry cleaning is cleaning without water. The clothes are tumbled in solvents like the trichloroethane mentioned above. This removes greasy dirt and many stains. The clothes are then spun and hung up to let the solvent evaporate.

Trichloroethane to the rescue!

1   Explain why water:
    a) is the most important solvent   b) is useful for washing
    c) can't wash away chewing gum. ▲
2   What happens in dry cleaning? ▲
3   Nail varnish has to be insoluble in water. Why?
4   Make a table headed:

Stains	Stain remover

Then work out the puzzle above and fill in the table.

5.4 Solutions without water

Many household substances are solutions. Here are a few which do not have water as the solvent.

Tincture of iodine is an antiseptic solution sometimes found in first aid boxes. It can be painted on to a cut to kill off germs.

The iodine is the germ killer, but iodine cannot be used on its own. It is a solid and cannot spread out well. Instead, a solution of iodine and alcohol is used. (Alcohol is used as the solvent because it dissolves iodine better than water.) When the solution is put on the cut, it carries the iodine into every crack in the skin. Then the alcohol evaporates, leaving the iodine to do its work.

Solvents are often used to spread out solutes in this way:

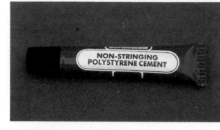

Varnish is painted on to wood as a solution. When the white spirit evaporates, the wood is left with a protective coating of solid resin.

Polystyrene glue solution is spread between two surfaces. The solvent quickly evaporates, leaving solid polystyrene which sticks the surfaces together.

Ball point ink is a solution with a very special mixture of solvents. These solvents have to keep the ink flowing freely, but must dry up quickly on the paper.

Perfume may contain a solvent to help to spread it out. The pure perfume is dissolved in a liquid like alcohol which evaporates quickly on the skin.

Did you know?

- 1 kg of rose petals produces only 1.2 g of pure perfume.
- Many glues give off very harmful fumes.

How perfume is extracted

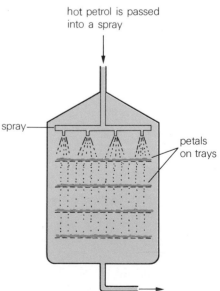

hot petrol is passed into a spray

spray

petals on trays

A mixture of perfume and petrol is run off. Then the petrol is removed

1 It is better to put iodine onto a cut as a solution, not as a solid. Explain why. ▲
2 What happens when polystyrene glue dries? ▲
3 What is special about the solvents used in a) ink ▲
   b) hair spray?
4 When do you use: a) an antiseptic b) varnish?
5 The diagram shows how pure perfume is extracted from roses using petrol. Explain what happens.
6 **Try to find out:** what the earliest inks were made of.

5.5　Water fit to drink

Rain water is fit to drink. So is most water in high mountain streams. But water flowing over the land may quickly:

- pick up small, insoluble particles of mud and dirt
- collect germs
- dissolve a little of the material it flows over.

The water in a big river is not pure. It is probably not even fit to drink. Most of our drinking water comes from big rivers and lakes. To make this water safe to drink, it is treated at the water works.

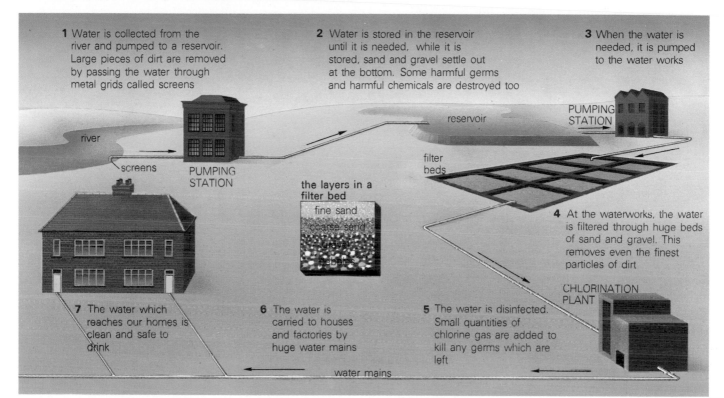

1 Water is collected from the river and pumped to a reservoir. Large pieces of dirt are removed by passing the water through metal grids called screens

2 Water is stored in the reservoir until it is needed, while it is stored, sand and gravel settle out at the bottom. Some harmful germs and harmful chemicals are destroyed too

3 When the water is needed, it is pumped to the water works

river
screens
PUMPING STATION
reservoir
filter beds
PUMPING STATION

the layers in a filter bed
fine sand
coarse sand
gravel
pebbles

4 At the waterworks, the water is filtered through huge beds of sand and gravel. This removes even the finest particles of dirt

CHLORINATION PLANT

7 The water which reaches our homes is clean and safe to drink

6 The water is carried to houses and factories by huge water mains

5 The water is disinfected. Small quantities of chlorine gas are added to kill any germs which are left

water mains

Did you know?

- In 1981, about 1 800 000 000 people (about half of the world's population) had no clean drinking water.
- 25 000 000 people die each year from drinking unclean water.

1 Why is river water less clean than the water in a mountain stream?

2 What are: a) screens　b) water mains　used for? ▲

3 What happens to water in　a) the reservoir　b) the filter beds? ▲

4 a) Why is chlorine added to our drinking water? ▲
b) Why must the amount of chlorine added be very small?
c) Why is extra chlorine added to the water in swimming pools?

5 If the water works break down, you should boil your drinking water. Why?

6 **Try to find out:** why fluoride is added to some water supplies.

rapid filter
(used in many waterworks instead of filter beds)

water is forced in under pressure

filter material

dirt is removed here

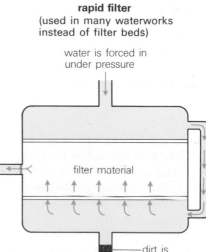

5.5 Purifying water

Carefully treated tap water is completely clear. It does not have any insoluble material floating in it. But it is not pure. When tap water is evaporated on a watch glass, a little white solid is left. Tap water has chemicals called **salts** dissolved in it.

You could produce pure water, even from salty sea water, using the **distillation** apparatus shown below. In distillation, the water is **evaporated**. Then the steam is **condensed**:

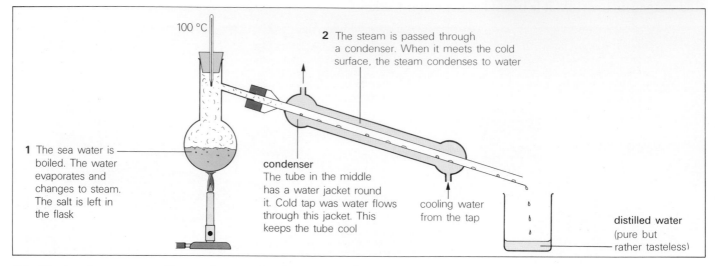

1 The sea water is boiled. The water evaporates and changes to steam. The salt is left in the flask

100 °C

2 The steam is passed through a condenser. When it meets the cold surface, the steam condenses to water

condenser
The tube in the middle has a water jacket round it. Cold tap was water flows through this jacket. This keeps the tube cool

cooling water from the tap

distilled water (pure but rather tasteless)

Distillation is not normally used to get drinking water from sea water. It is very expensive because it uses up so much energy. It is only used on a large scale in countries like Saudi Arabia where water is scarce and energy is cheap.

Many ships, however, carry distillation apparatus. Some cruise ships use it to produce all their drinking water from sea water. Other ships carry it to produce emergency drinking water. In the 1973 Round the World Yacht Race, Chay Blyth and his crew had to invent their own apparatus when they lost drinking water from a leaking tank.

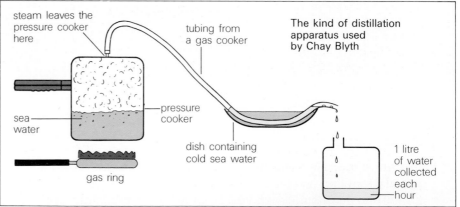

steam leaves the pressure cooker here

tubing from a gas cooker

The kind of distillation apparatus used by Chay Blyth

sea water

pressure cooker

dish containing cold sea water

gas ring

1 litre of water collected each hour

1 How can you show that tap water is not pure? ▲
2 Complete the sentence: 'In distillation, water is . . ., then . . .'. ▲
3 a) Why does distillation produce pure water from sea water?
   b) Where is distillation used to produce drinking water from sea water, and why?
4 Distilled water is tasteless. Sea water has too much taste. Why?
5 Explain how Chay Blyth's distillation apparatus worked.
6 **Try to find out:** where the Dead Sea is and why it is so salty.

Did you know?

- In the liner Queen Elizabeth II, one million litres of water are distilled every day.
- Distilled water should be used in car batteries and steam irons.

Britain's waterworks purify huge quantities of water. Every day, they supply about 20 000 000 000 litres of purified water.

Who uses all the water?

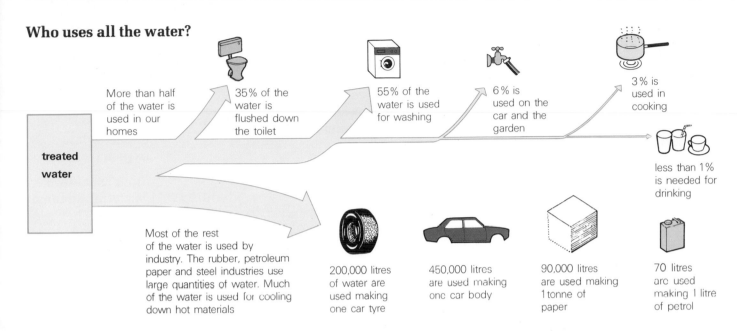

More than half of the water is used in our homes

treated water

35% of the water is flushed down the toilet

55% of the water is used for washing

6% is used on the car and the garden

3% is used in cooking

less than 1% is needed for drinking

Most of the rest of the water is used by industry. The rubber, petroleum paper and steel industries use large quantities of water. Much of the water is used for cooling down hot materials

200,000 litres of water are used making one car tyre

450,000 litres are used making one car body

90,000 litres are used making 1 tonne of paper

70 litres are used making 1 litre of petrol

Could some of the water be saved?

Because the British climate is usually so wet, there does not seem to be much need to try to save water. But, if more water was saved, fewer reservoirs would have to be built. If fewer reservoirs were needed, money would be saved, and fewer beautiful mountain valleys would have to be flooded.

Water can be saved in different ways.
We can save water at home by mending dripping taps, by using hosepipes less and by turning off running taps.
Industry can save water by **recycling** much of the water it uses. (To 'recycle' means to 'use again'.)
Water Boards could save more than anyone by replacing or mending leaking pipes. On average, about one quarter of the treated water leaks away before it reaches the user.

The valley has been flooded to make a reservoir

Did you know?

- Most washing machines use up to 180 litres of water every wash.
- You probably flush 40 litres of water down the toilet every day.

1　What is most water used for:　a) at home　b) by industry? ▲
2　How can water be saved:　a) at home　b) by industry? ▲
3　What is meant by 'recycling'? ▲
4　Draw a bar chart to show how water is used in the home.
5　**Try to find out:** whether your water is supplied from a man made lake.

5.5 Cleaning up a river

Bad News

Rivers become polluted whenever anything harmful is poured into them. Some rivers are so badly polluted that they contain no living creatures at all. River animals are sometimes killed by poisonous chemicals which are dumped in the river by industry. But they can also be killed by more everyday and less poisonous materials like fertilisers, detergents, sewage and even hot water. These materials pollute the water by reducing the amount of oxygen which is dissolved in it. River animals need oxygen to live. The smaller the amount of oxygen dissolved in the river water, the smaller is the number of animals which can live in the river.

Good News

Polluted rivers can be cleaned up. In 1951, the River Thames was very badly polluted. But as you can see from the dissolved oxygen levels on the map below, the river is much cleaner now. The main steps in the clean-up were:

1 reducing the amount of industrial waste in the river
2 persuading soap manufacturers only to make detergents which can be broken down at the sewage works
3 improving sewage works and building new ones.

This last step was the most important – about 80% of the pollution was due to untreated sewage.

pollutant	how it affects the dissolved oxygen
detergent	Water with detergent in it dissolves less oxygen
hot water	Hot water dissolves less oxygen than cold.
fertilisers raw sewage	Both encourage bacteria and fungi to grow. They use up oxygen.

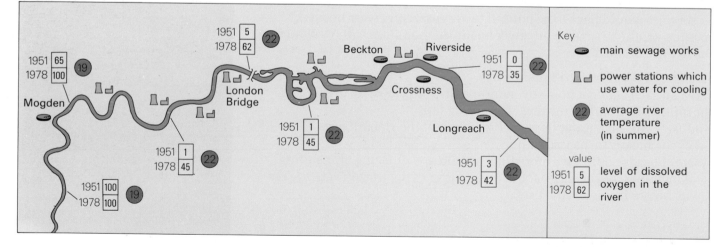

1 When do rivers become polluted? ▲
2 Why is the amount of oxygen dissolved in river water affected by
   a) untreated sewage b) detergents c) hot water? ▲
   About the Thames:
3 Compare the dissolved oxygen levels of 1978 and 1951 and explain the differences.
4 Where is dissolved oxygen: a) highest b) lowest? Why?
5 Why does the water temperature change as it flows down river?
6 **Try to find out:** about any other kind of water pollution.

Did you know?

- In 1951, a 40 km stretch of the Thames had no dissolved oxygen *at all*.
- 97 different types of fish have been found in the Thames since it was cleaned up.

64

Cells and reproduction

The **microscope** is a useful
scientific instrument. When you
look at something through a
microscope, it appears larger –
the microscope **magnifies** what
you see.

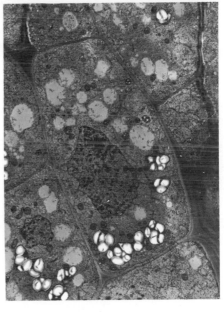

If you look at a plant root through
a low power microscope, this is
what you see. The root looks a bit
like a brick wall.

With a more powerful
microscope, you can see that each
'brick' is made up of a number of
different parts.

With a really powerful
microscope, you can see even the
smallest parts very clearly.

Each 'brick' in the plant root is a **cell**.
Cells are the 'building blocks of life'. Every living thing is made up of
one or more cells. Most animals and plants are made up of millions of
cells joined together.

Scientists have discovered lots of things about cells by using
microscopes. They can explain how cells work, how new cells are
produced, and how special cells allow new animals to grow. That's
what this section is about.

6.1 Cells

Before the photographs of the cells shown below were taken, a dye was put on the cells. This helps to show up the different parts of each cell. The photographs were then taken with a camera fitted to a microscope:

Cheek cell

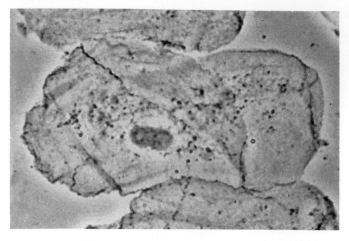

This is a photo of a cheek cell. The inside of your cheek is made up of many of these cells. In fact, your whole body is made up of millions and millions of tiny living cells joined together.

A diagram to help you see the main parts of the cheek cell is given below:

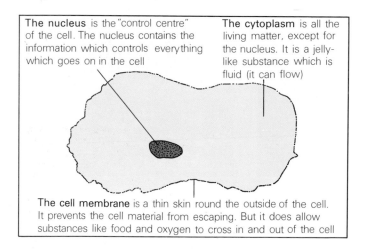

The nucleus is the "control centre" of the cell. The nucleus contains the information which controls everything which goes on in the cell

The cytoplasm is all the living matter, except for the nucleus. It is a jelly-like substance which is fluid (it can flow)

The cell membrane is a thin skin round the outside of the cell. It prevents the cell material from escaping. But it does allow substances like food and oxygen to cross in and out of the cell

Pond weed cell

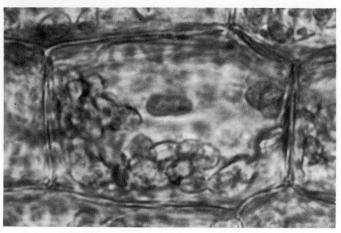

This is a photo of a cell from pond weed. The weed is made up of millions of tiny living cells, too.

A diagram to help you see the main parts of the pond weed cell is given below:

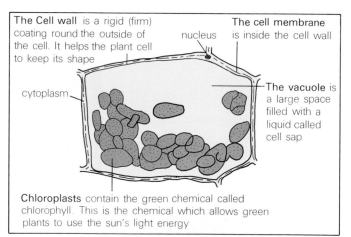

The Cell wall is a rigid (firm) coating round the outside of the cell. It helps the plant cell to keep its shape

nucleus

The cell membrane is inside the cell wall

cytoplasm

The vacuole is a large space filled with a liquid called cell sap

Chloroplasts contain the green chemical called chlorophyll. This is the chemical which allows green plants to use the sun's light energy

1 Why is the nucleus called the 'control centre' of the cell? ▲
2 What is the cytoplasm? What is it like? ▲
3 What is a cell vacuole? ▲
4 a) What do chloroplasts contain? ▲
  b) Why is this important? (Look at page 27 if you need help!)
5 Pick out the three parts which are found both in the cheek cell and in the pond weed cell (and in most other cells, too!).
6 **Try to find out:** when cells were discovered, and by whom.

Did you know?

● The cell membrane is only 0.000 01 mm thick.
● Animal cells don't have chloroplasts. They don't have cell walls either.

6.1 Single-celled organisms

You need a microscope to see the smallest living organisms. They are made up of only one cell. That cell is an 'all purpose' cell. It is able to take in food, to grow and reproduce, to take in and get rid of gases, to react, to produce energy and to move.

Amoeba is a one-celled animal which lives in water. It has no fixed shape. It is like a bag of jelly. It moves by pushing out a 'foot' and flowing into it. It catches its food in this way, too.

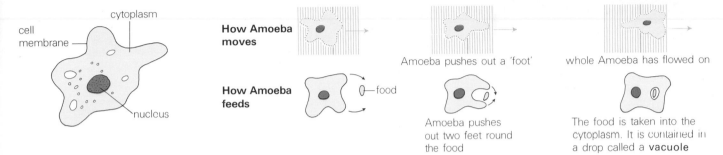

cytoplasm

cell membrane

nucleus

How Amoeba moves

Amoeba pushes out a 'foot'

whole Amoeba has flowed on

How Amoeba feeds

food

Amoeba pushes out two feet round the food

The food is taken into the cytoplasm. It is contained in a drop called a **vacuole**

Binary Fission

Amoeba **reproduces** by dividing in two. This method of reproduction is called **binary fission**. (fission = dividing, binary = two part)
Many single-celled organisms reproduce in this way.

This is how binary fission takes place:

1 The information material in the cell's nucleus is doubled

2 The nucleus divides. Each half contains the same information. The halves go to the opposite ends of the cell

3 The cytoplasm divides in two

4 Two new daughter cells' are produced

Each of the two **daughter cells** is an exact copy of its parent. Each carries the same information. When a daughter cell has grown to be full size, it divides, too.

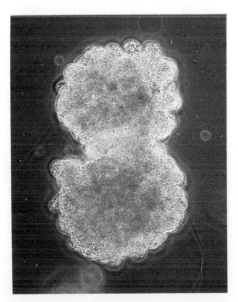

Did you know?

● Amoeba can be found in stagnant ponds and ditches.
● Amoeba moves away from strong light.

1 What is: a) an organism (see page 13) b) reproduction (see page 13) c) binary fission? ▲
2 How is the information passed on when a cell divides?
3 Why can Amoeba be described as an 'all purpose cell'?
4 Give two examples of how Amoeba can react to things which happen round about it.
5 **Try to find out:** about other single-celled organisms.

6.1 Special cells for special purposes

Amoeba is one of the simplest living organisms. You are one of the most complicated.

Your body is made up of a number of different **organs** such as the heart, lungs, eyes, brain, liver and tongue. Each organ has a particular job to do.

An organ is made up of **tissues** which work together. Skin, muscle and bone are three of the body's tissues.

Each tissue is made up of a large number of similar cells grouped together. These cells all work together to do a particular job.

There are many different kinds of cell in the body. Each has a special job to do. The diagrams below show a few of these different cells – there are lots more. In your body there are:

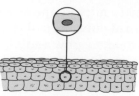

skin tissue a collection of skin cells

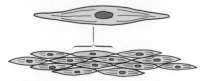

muscle tissue a collection of muscle cells

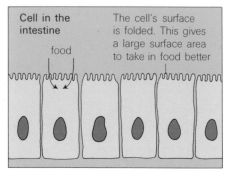

1 cells which take in or **absorb** food.

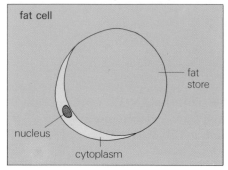

2 cells which store food.

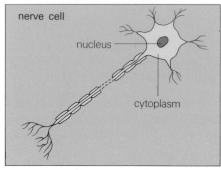

3 cells which carry messages round the body.

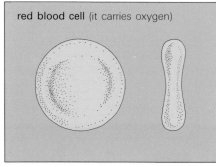

4 cells which carry chemicals round your body.

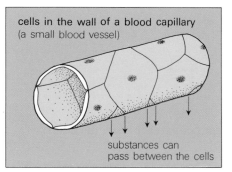

5 cells which allow chemicals to pass through channels between them.

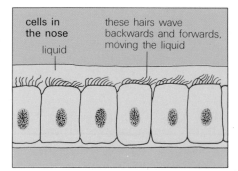

6 cells with tiny hairs which move liquids over their surface.

1 a) Name some of the organs in your body.
   b) What are organs made up of? ▲
2 What is special about the cells in a tissue? ▲
3 What is a capillary? ▲
4 How is a red blood cell different from most other cells?
5 Why is it important that: a) an intestine cell has a folded surface b) capillary cells have channels between them?
6 Why do fat cells shrink when you slim?
7 **Try to find out:** how your nose helps to clean the air you breathe in.

Did you know?

- Red blood cells have no nucleus.
- Red blood cells are bigger than the smallest capillaries. They are squeezed out of shape as they pass through the capillaries.

Only the simplest animals reproduce by dividing in two. Most animals, including humans, produce special **sex cells** for reproduction.

There are usually two types of sex cell. The smaller of the two can move on its own. It is called the **male cell** or **sperm**. The larger one cannot move on its own. It is called the **female cell** or **egg**.

Fertilisation

Fertilisation has to take place before a new animal can grow. In fertilisation, the sperm and the egg join up, making a **fertilised egg**. The new animal grows from this fertilised egg which divides, producing more and more cells.

Sexual reproduction

Reproduction of the type which begins when a sperm joins with an egg is called **sexual reproduction**. Most animals reproduce by sexual reproduction involving two parents. Each parent has special **reproductive organs** which produce the sex cells. The male has **testes** which produce the sperms. The female has **ovaries** which produce the eggs. The number of sperm cells produced by the male is far greater than the number of egg cells produced by the female.

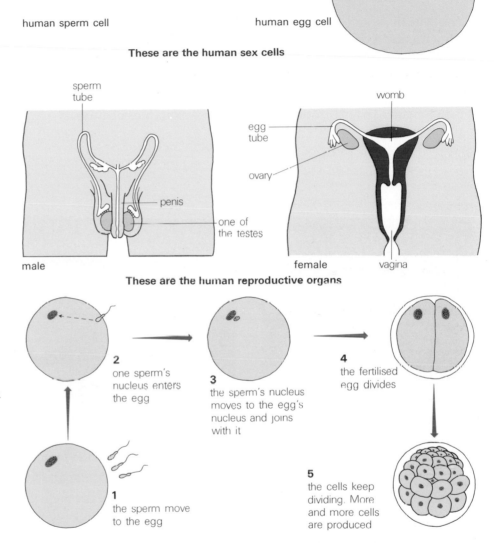

These are the human sex cells

These are the human reproductive organs

This is what happens when an egg is fertilised

1 What are male sex cells called? Where are they produced? ▲
2 What are female sex cells called? Where are they produced? ▲
3 a) What happens when an egg is fertilised?
   b) What happens to the egg after fertilisation? ▲
4 What is meant by sexual reproduction? ▲
5 a) Sperm cells, the male cells, are described as many, minute and mobile. Explain why.
   b) What are the differences between egg cells and sperm cells?
6 **Try to find out**: what is meant by *asexual reproduction*.

Did you know?

- A human egg is slightly smaller than this →. A human sperm is many times smaller.
- A snail can produce both sperms and eggs. So can an earthworm and many other animals.

6.2 Sperm meets egg

'How does the sperm get to the egg?'
There is really only one answer to that question. The sperm gets to the egg by *swimming*! But in some cases, the sperm meets the egg outside the mother's body (**external fertilisation**). In other cases, the sperm and egg meet inside the mother's body (**internal fertilisation**).

In the case of most fish and amphibians, external fertilisation takes place. The male and female release their sperm and eggs into the water. If a sperm meets an egg, fertilisation takes place. If they don't meet, both sperm and eggs die. In fact, many of the eggs produced by the female are not fertilised.

Many animals pair up before the sperm and eggs are released. This gives a better chance of fertilisation. Here are three examples of animals pairing up:

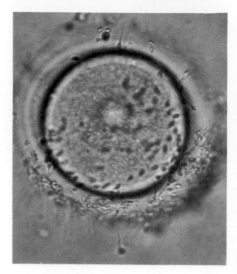

Sperm meets egg

1 Many fish swim side by side when releasing sperm and eggs into the water.

2 Sticklebacks make a nest where the female lays her eggs. The male then covers the eggs with sperm.

3 The male frog sits on the female's back. He produces sperm while she is laying her eggs.

In the case of reptiles, birds and mammals, internal fertilisation takes place. These animals have reproductive organs which allow the male to pass sperm into the female's body. There, the sperm travel along tubes which lead to the egg. They swim through liquid which these tubes produce.

1   What is the difference between internal fertilisation and external fertilisation? ▲
2   Which sex cell moves by itself, the sperm or the egg? How does it move? ▲
3   Why is pairing important for animals whose eggs are fertilised outside the female's body?
4   Which method of fertilisation gives the egg the best chance of being fertilised? Explain your choice.
5   **Try to find out:** how fish eggs are fertilised on a fish farm.

Did you know?

- Sperm were first noticed in the 17th century – using a microscope.
- The sperm's 'tail' lashes backwards and forwards like a whip. That's what drives the sperm forward.

6.2   Passing on the instructions

It's hard to believe that a cell's nucleus contains all the instructions needed for the cell to do its work. It's even harder to believe that the nucleus of a fertilised egg contains all the instructions needed for a new animal to grow. But it does – and the passing on of these instructions is all-important.

Chromosomes

The instructions are carried on fine threads of material called **chromosomes**. Every cell in the animal's body, apart from sperm and egg cells, has the same number and kind of chromosomes in its nucleus. Human cells have 46 chromosomes, horse cells have 60, mouse cells have 40 and so on. Sperm and eggs have only half that number. When a sperm and an egg join up, a new animal begins to grow. Fertilisation is really the putting together of a 'full set' of instructions needed for the animal to grow. Half of the instructions are passed on from the father, in the sperm, half from the mother, in the egg.

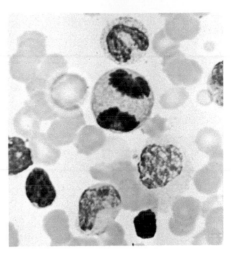

Chromosomes dividing in human blood cells

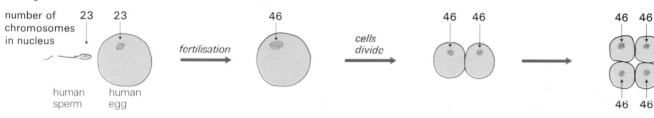

As more and more cells are produced and the animal develops, the same instructions are passed on from one cell to the next. This is how each cell gets the same set of chromosomes when a cell divides. To make things simpler, only 4 chromosomes are shown in the diagram. There are three main stages:

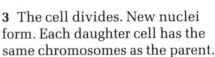

1 Each chromosome makes an exact copy of itself.

2 The copies separate. One chromosome from each pair goes to opposite sides of the cell.

3 The cell divides. New nuclei form. Each daughter cell has the same chromosomes as the parent.

1   What is a chromosome? Where are chromosomes found? Why are they important? ▲
2   How many chromosomes are there in the nucleus of: a) a human skin cell   b) a horse sperm cell   c) a mouse egg cell?
3   When a human egg is fertilised: 23 + 23 = 46.
    When a human cell divides: 46 → 92 → 46 + 46.
    In each case, explain what is meant.
4   **Try to find out:** what the 'instructions' on a chromosome are called.

Did you know?

- A chromosome is only about 0.005 mm long.
- Scientists are now able to change some of the chromosome material in a nucleus.

6.3   The next generation of humans [1]

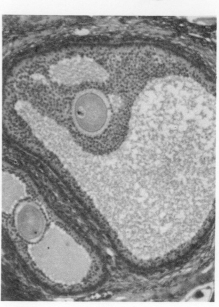

(Look back at the diagrams on page 69 before reading on.)

Ovulation

Inside every girl's ovaries, there are many partly-formed eggs – perhaps one million of them. None of these eggs will form completely in the first part of her life. But when she is a woman, a fully developed egg is released from one of her ovaries roughly once every four weeks, and starts to travel down the egg tubes. This process of egg release is called **ovulation**.

Sexual intercourse

If a man and woman want to have a child, sperm from the man has to reach an egg while it is inside the woman's egg tubes. For this to happen, the man's penis is slipped into the woman's vagina, and sperm cells are released. This is called **sexual intercourse**. The sperms then travel through the womb to the egg tubes. If one of the sperms meets the egg, fertilisation takes place. The fertilised egg travels down the egg tube and becomes fixed to the new, soft lining of the womb.

Partly-formed egg cells

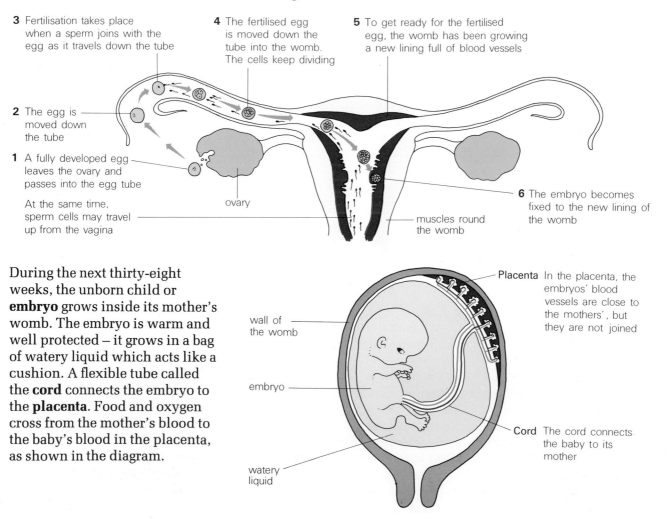

3 Fertilisation takes place when a sperm joins with the egg as it travels down the tube

4 The fertilised egg is moved down the tube into the womb. The cells keep dividing

5 To get ready for the fertilised egg, the womb has been growing a new lining full of blood vessels

2 The egg is moved down the tube

1 A fully developed egg leaves the ovary and passes into the egg tube

At the same time, sperm cells may travel up from the vagina

ovary

muscles round the womb

6 The embryo becomes fixed to the new lining of the womb

During the next thirty-eight weeks, the unborn child or **embryo** grows inside its mother's womb. The embryo is warm and well protected – it grows in a bag of watery liquid which acts like a cushion. A flexible tube called the **cord** connects the embryo to the **placenta**. Food and oxygen cross from the mother's blood to the baby's blood in the placenta, as shown in the diagram.

wall of the womb

embryo

watery liquid

Placenta In the placenta, the embryos' blood vessels are close to the mothers', but they are not joined

Cord The cord connects the baby to its mother

The embryo grows as more and more cells are produced. Gradually, it begins to take up the shape of a human. After it has been growing for about eight weeks, it has legs, arms, eyes, ears and a mouth. Within three months, *all* the different types of cell for a complete, new human being have been made. The embryo is then called a **foetus**.

Birth

After about nine months of growth, the baby is ready to be born, and the mother 'goes into labour'. In labour, the muscles in the womb tighten or **contract**. The opening in the womb becomes larger. The bag of fluid in which the baby has been held bursts.
At first, the muscles may contract once every half hour. But soon the contractions come more often and are stronger. Gradually the muscles push the baby out of the womb, through the vagina and into the world outside.

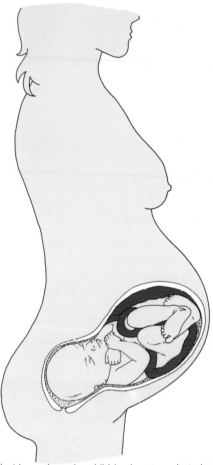

Inside mother - the child is almost ready to be born

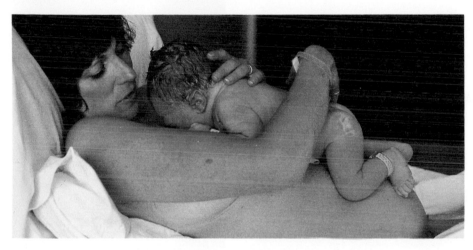

◄ The mother greets her new-born child with a quiet cuddle

If the egg is not fertilised . . .

Most of the eggs released from the ovaries are not fertilised. An egg dies if it has not been fertilised 48 hours after being released. After this time, a big change takes place in the new lining of the womb which had been growing, ready to receive an embryo. The new lining is no longer needed – there is no growing embryo for it to support. So it is shed by the womb. It leaves the body through the vagina, producing the monthly bleeding called a **period** or **menstruation**.

1   Where is an egg: a) produced   b) fertilised? ▲
2   When the baby is in the womb: a) how does it get food and oxygen   b) how is it protected? ▲
3   What are contractions? How is a baby born? ▲
4   a) Why does the growing embryo depend on the cord?
    b) Why can the cord be safely cut once the baby is born?
5   If a woman begins to release eggs at the age of 15, and stops releasing them at 45, how many eggs will she produce?
6   **Try to find out:** what special foods a mother-to-be should have in her diet, and why.

Did you know?

• A sperm can only move between 1 and 3 mm a minute.
• 300 million sperm cells may be released during intercourse. However, only about 100 will reach the egg and only *one* will join with it.

6.3 Baby: the first nine months

We are used to thinking of a baby being age 0 on the day it is born. But it has been growing for about nine months before it enters the world. This is what happens in the nine months (38 weeks) before it is born:

Week

0 Egg is fertilised in the tube.

1 Embryo becomes attached to the womb.

2 Embryo's eyes begin to develop. Its legs and arms are tiny bumps.

6 Embryo begins to look like a human. Ears, hands and feet begin to grow. Heart begins to beat.

10 Baby's fingers and toes grow. It can move its arms and legs a little – it can even swallow and frown!

14 If doctors could see the baby, they could tell if it was a boy or girl.

18 Baby has hair, eyebrows. Doctors can hear its heart beat. It can move . . . mother begins to feel its kicks!

26 Baby opens its eyes.

30 If born now, the baby could live with special care.

34 Baby has grown a lot of fat in the last four weeks, to keep it warm when it is born.

38 Baby is born.

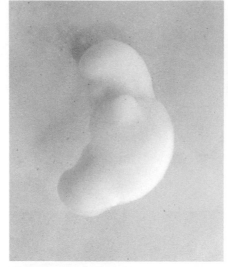

Week 2   1.5 cm   7 g

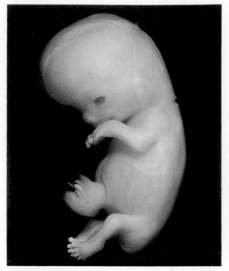

Week 6   2.5 cm   20 g

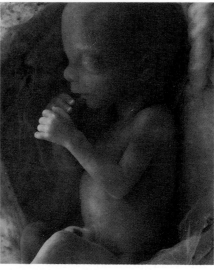

Week 12   12 cm   50 g

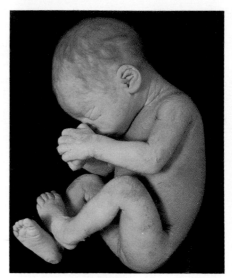

Week 38   Happy birthday !

1 When does a baby: a) begin to move   b) open its eyes   c) begin to kick its mother so that she notices?   ▲

2 When is the baby: a) as long as a finger nail   b) as long as your foot?

3 Babies which are born early are kept warm in incubators. Why?

4 Draw two graphs, one to show how the baby's length increases and one to show how its weight increases as it grows. (**More information**: Week 16: 15 cm, 170 g; Week 24: 30 cm, 670 g; Week 32: 40 cm, 1600 g.)

5 **Try to find out:** why girls should be given injections to prevent them from getting German measles.

Did you know?

- The baby swallows liquid when it is in the womb. It practises breathing and sucking, too.
- The Chinese consider a baby to be one year old on the day it is born.

6.3 Genes

(Read page 71 again before you go on.)

Whether you are naturally tall or small, blond- or brown-haired, blue- or green-eyed, left- or right-handed depends largely on the **genes** in your body cells. A gene is part of one of the chromosome threads inside the nucleus of each cell. Genes control the characteristics which are passed on from parents to their children.

When an egg is fertilised, two sets of genes are put together. One set comes from the father in the sperm's chromosomes. The other set, from the mother, is in the egg's chromosomes. Each set carries half of the total instructions for the new human.

The two sets of genes in a fertilised egg are very similar. The genes passed on by the father for example, carry instructions about body characteristics like height, hair colour and eye colour. The genes passed on by the mother carry instructions about these characteristics, too.

Dominant genes

The genes in one set may be exactly the same as the genes in the other set. Then the results are what you would expect. If each parent passes on a brown eye gene to a child, that child will have brown eyes. But the genes in the two sets may be different. Then the genes with the 'stronger', or **dominant** instructions will control what happens. The instructions carried by a brown eye gene are 'stronger' than the instructions carried by a blue eye gene. If one parent passes on a brown eye gene and the other passes on a blue eye gene, the child will have brown eyes.

Heredity

The passing on of characteristics from one generation to the next is called **heredity**. It's a complicated business. For one thing, there are up to 10 000 genes on each chromosome. For another, the genes passed on by a father (or mother) to one child will probably be different from the genes passed on to another child. It's doubtful whether scientists will ever be able to say exactly which characteristics will be passed on from parent to child. They will probably never be able to give a completely accurate answer to the question 'What will the baby look like?'.

1 What is meant by: a) a gene   b) heredity? ▲
2 What kinds of instructions are carried on the genes? ▲
3 How are genes passed on from father and mother to child? ▲
4 If a father passes on a black hair gene and the mother passes on a blond hair gene, the child will have black hair. What does this tell you about blond and black hair genes?
5 Mr. and Mrs. Smith, as shown on the right, each have one brown eye gene and one blue eye gene. The arrows show which gene is passed to each child.
   a) Why do Mr. and Mrs. Smith have brown eyes?
   b) What colour eyes do Julie, John and Joan have?

Did you know?

- You can't teach yourself how to roll your tongue. Whether you are a tongue roller or not depends on the genes you have inherited!

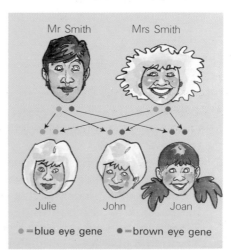

- Sometimes, the ball of cells which grows from the fertilised egg splits into two. Each half grows into a baby. That's how identical twins are produced, and that's why identical twins have identical genes.

Mr Smith   Mrs Smith

Julie   John   Joan

●=blue eye gene   ●=brown eye gene

6.4 Growing up: non-human animals

Embryos of different animals have many things in common:

- They all grow from fertilised eggs which divide, producing more and more cells.
- They all need food and oxygen to grow.
- They are all surrounded by water as they grow.

But different embryos grow in very different ways:

- Some embryos grow in eggs which the mother has laid. Others grow inside the mother's body.
- Some embryos get all their food from the egg's yolk. The oxygen they need travels to them through the egg. Others get food and oxygen from the mother's blood.
- Most embryos grow into 'young adults'. But the embryos of insects and some amphibians grow into **larvae** (like caterpillars, maggots and tadpoles). The larvae change into adults later.

You have already read about the human embryo, which grows inside its mother's womb. Here are two embryos which grow inside eggs until they have completely developed:

A newly-hatched chicken

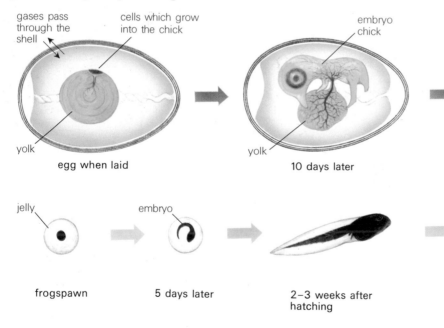

gases pass through the shell

cells which grow into the chick

yolk

egg when laid

embryo chick

yolk

10 days later

young adult

jelly

frogspawn

embryo

5 days later

2–3 weeks after hatching

1 Which two things do all growing embryos need? ▲
2 How does: a) a frog embryo b) a human embryo get the oxygen it needs?
3 a) What is the yolk's job in the egg? ▲
  b) As the chicken grows, the egg's yolk gets smaller. Why?
4 Make lists of animals which grow: a) inside the mother
  b) inside eggs which the mother lays.
5 Which animals produce embryos that grow into larvae? ▲
6 **Try to find out:** how lizard embryos grow.

Did you know?

- The time from the fertilising of the egg to the birth of the young adult is called the **gestation period**. The gestation period is about:
  21 days for a chicken
  61 days for a cat
  267 days for a human
  645 days for an elephant.

6.4 Looking after the young ones

Young animals grow up in a dangerous world. They can't all rely on their parents for care and protection.

These animals get little or no protection from their parents

Most young fish, amphibians and reptiles grow up with little or no care from their parents. The parents produce fertilised eggs, then leave them to develop on their own. Many of the young die or get eaten. A few parents (mostly reptiles) protect their eggs by laying them in nests or in underground burrows. Only a very few, like the African toad, the stickleback and the crocodile, actually guard the eggs and young.

Young birds and mammals need, and usually get, a lot of care to help them survive. Bird and mammal families are small, which makes it easier to care for the young. Bird parents protect their eggs and work hard to feed and protect the young birds when they hatch. Young mammals are fed by their mother and protected by their parents, too.

These animals are well cared for

Perhaps the 'most devoted parent' award should go to the male Emperor Penguin. His mate produces one egg, and he looks after it while she goes off to feed. He balances the egg on his feet and tucks it into his feathers and skin to keep it from freezing. Then he stands, huddled together with other fathers, trying to keep warm. He may do this for up to one month at a time in temperatures as cold as −40°C. If his mate spends a month feeding, he will spend that month starving!

Did you know?

- Only 0.1% of all animals care for their young.
- Some fish, called mouthbreeders, take their eggs into their mouths and keep them there till they hatch.

This animal gives 5-star care!

1 What quality of care is given to their young by: a) birds b) most fish c) the stickleback d) mammals? ▲
2 Why must a penguin's egg not be left on the ground?
3 Why must birds sit on their eggs until they hatch?
4 What food does a young mammal get from its mother?
5 List some of the dangers which face young animals as they grow.
6 **Try to find out:** how the cuckoo makes sure that its young are well looked after.

6.4 The struggle for survival

A species of animal has a **stable population** if its numbers stay roughly the same over several years. All species which have stable populations have one thing in common: enough young survive to replace the adults which die off or are killed.

The cod, the common frog, the robin and the red deer are four 'successful' British species. In areas where man does not interfere with them, they are in little danger of dying out. Their populations are more or less stable even though the different animals reproduce in very different ways.

In the struggle for survival, their young don't all have an equal chance, and the number of young produced varies enormously, as the table shows. You will need information from it to answer the questions:

The red deer has lived in Britain for thousands of years

	cod	common frog	robin	red deer
number of eggs produced per year:	6 million	2000–3000	6 (twice a year)	usually 1
where the eggs are fertilised	in water (in the sea)	in water (in ponds and ditches)	inside the mother	inside the mother
care and protection given to the egg and the young animal	None. The parents leave the egg floating in the water. The young have to hunt for their own food.	None from the parents, but the egg is covered with a foul-tasting jelly. The tadpoles find their own food.	The parents sit on the eggs till they hatch. Then they feed and protect the young.	The mother feeds the fawn with milk for the first 8 months. She protects it for up to 1½ years.
enemies which prey on the eggs or the young	most other fish (even cod!)	fish, newts, water insects, herons, hedgehogs	crows, hawks, squirrels	very few. Occasionally foxes and eagles attack weak fawns.

1 What is meant by saying 'The red deer population is stable'? ▲
2 Which of the four species described above gives least care and protection to its young? Why does this species not die out?
3 What connection can you see between the number of eggs produced by a species and the care and protection given to its young?
4 In many parts of the country, farmers drain marshes to get extra land. In these areas, the frog population is falling. Suggest why.
5 When will an animal species be in danger of becoming extinct?
6 **Try to find out:** how and why the British osprey population has changed in the last 80 years.

Did you know?

- Only 3 out of every 100 000 fertilised mackerel eggs grow to be adult fish.
- The normal lifespan of a British robin is only 5–6 months.

6.5 Reproduction in flowering plants

Every part of a plant has its own special job to do. The flower's job is to allow the plant to reproduce by sexual reproduction. It produces the male sex cells (which are contained in **pollen grains**) and the female sex cells (contained in the **ovules**).

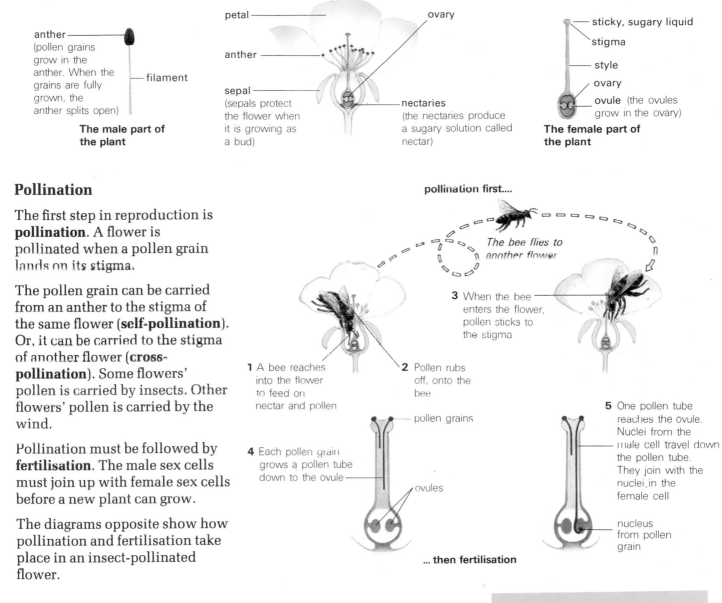

anther
(pollen grains
grow in the
anther. When the
grains are fully
grown, the
anther splits open)

filament

**The male part of
the plant**

petal

ovary

anther

sepal
(sepals protect
the flower when
it is growing as
a bud)

nectaries
(the nectaries produce
a sugary solution called
nectar)

sticky, sugary liquid

stigma

style

ovary

ovule (the ovules
grow in the ovary)

**The female part of
the plant**

Pollination

The first step in reproduction is **pollination**. A flower is pollinated when a pollen grain lands on its stigma.

The pollen grain can be carried from an anther to the stigma of the same flower (**self-pollination**). Or, it can be carried to the stigma of another flower (**cross-pollination**). Some flowers' pollen is carried by insects. Other flowers' pollen is carried by the wind.

Pollination must be followed by **fertilisation**. The male sex cells must join up with female sex cells before a new plant can grow.

The diagrams opposite show how pollination and fertilisation take place in an insect-pollinated flower.

pollination first....

The bee flies to
another flower

3 When the bee
enters the flower,
pollen sticks to
the stigma

1 A bee reaches
into the flower
to feed on
nectar and pollen

2 Pollen rubs
off, onto the
bee

pollen grains

4 Each pollen grain
grows a pollen tube
down to the ovule

ovules

5 One pollen tube
reaches the ovule.
Nuclei from the
male cell travel down
the pollen tube.
They join with the
nuclei in the
female cell

nucleus
from pollen
grain

... then fertilisation

1 Why are: a) sepals b) anthers c) ovaries important to a flower? ▲
2 What happens in: a) pollination b) fertilisation? ▲
3 Is the flower in the 5-stage diagram above being self-pollinated or cross-pollinated? Explain your answer.
4 How can pollen be carried from flower to flower? ▲
5 Pollen grain **A** is small and very light. Pollen grain **B** is sticky. One is pollen from a buttercup. The other is pollen from a grass. Say which is which, giving reasons for your answer.

Did you know?

- Bright flowers are usually insect-pollinated.
  Grass flowers are mostly wind-pollinated.
- If a pollen grain lands on the stigma of another kind of flower, the pollen grain dies.

6.5 How a new plant grows

After the ovules have been fertilised, most of the flower withers and dies. At the same time, the fertilised ovules grow inside the ovary until they develop into seeds. Each seed contains a tiny embryo plant. It also contains a food store. Round the seed is a tough **seed coat** which protects it.

The ovary grows, too. The developed ovary, with the seeds inside it, is called a **fruit**. A pea pod, a plum and a sycamore propeller are all fruits.

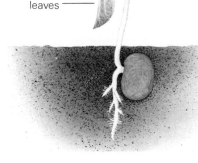

A broad bean seed
(cut across)

embryo shoot — tough seed coat

embryo root — food store

Scattering the seeds

In many plants, the seeds are scattered. Often the whole fruit is scattered, with the seed inside. This scattering is important. Plants which are overcrowded don't grow well. Scattering gives the plants more room to grow.

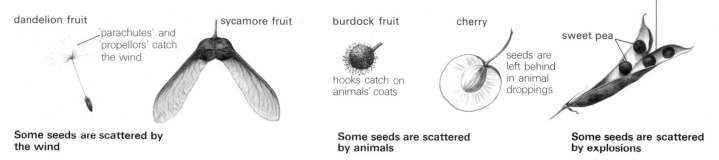

dandelion fruit

'parachutes' and 'propellors' catch the wind

sycamore fruit

burdock fruit

hooks catch on animals' coats

cherry

seeds are left behind in animal droppings

On a hot day, when the pod is dry, it bursts open. This scatters the seed.

sweet pea

Some seeds are scattered by the wind

Some seeds are scattered by animals

Some seeds are scattered by explosions

Germination

If the conditions are right, a seed will begin to grow or **germinate**. The seed which is scattered from the plant has very little water in it. A new plant begins to grow when the seed takes in water.

new leaves

new root

root hairs

1 When the bean seed takes in water, the root begins to grow, using the stored food for energy.

2 The root continues to grow. The root hairs begin to take in water from the soil.

3 The shoot begins to grow. Then leaves appear.

1. What happens to: a) the flower b) the ovule c) the ovary wall after fertilisation takes place? ▲
2. a) Give one reason why it is better for seeds to be scattered from the plant.
   b) Give three ways in which seeds are scattered. ▲
3. What happens when a seed germinates? Where does the growing embryo get its energy from? ▲
4. Seeds don't germinate in the packet. Why is this?
5. **Try to find out:** about other seeds, and how they are scattered.

Did you know?

- A strawberry isn't a real fruit. The fruits are the 'pips' on the outside.
- In 1954, 10 000 year old seeds were found in frozen ground in Canada. In 1966, they germinated!

Insects are the only animals which pollinate European flowers. But in other parts of the world, some flowers are pollinated by birds and bats which feed on the flowers' nectar and pollen in the same way that insects do.

All these animals are attracted by the flowers' 'advertising signs' – brightly coloured petals and strong scents. But different animals are attracted to different 'advertising signs', sometimes for the strangest reasons.

Birds can see red well. Bird-pollinated flowers are usually bright red, or orange, or scarlet.

Insects are attracted to most colours, but only butterflies and moths can see the red colour which we see.

Many flowers have coloured stripes called 'honey guides' on their petals. Once the animal arrives on the flower, the stripes guide it to the nectar.

Most insects are attracted to sweet smelling flowers. It is thought that the scents remind them of their food.

Carrion flowers are anything but sweet smelling! They smell, and look like rotting flesh. They are pollinated by insects which lay their eggs on dead flesh.

This orchid has such a strong scent that it makes any visiting bee 'drunk'. The bee falls into the flower. The only way it can escape is past the orchid's anthers.

1 Why do animals visit flowers? ▲
2 Why are insects attracted to sweet smelling flowers? ▲
3 Why are the insects which pollinate carrion flowers unlikely to pollinate sweet smelling flowers?
4 Would you expect bats to be more attracted by colour or scent? Explain.
5 'Red flowers in Britain are likely to be pollinated by moths and butterflies.' Explain why this is so.
6 **Try to find out:** the names of some perfumes made from flowers.

Did you know?

- Honeysuckle is pollinated by moths. It produces more scent in the evening than in the daytime.
- Bat-pollinated flowers produce huge amounts of pollen, enough to feed the bat and to pollinate the flowers.

What did you learn?

Here is a word search. It contains twenty words which you have read about in this section.

Write down the numbers **1** to **20** in your books. Then find the words. There is a clue to help you with each one. (Each word is only used once.)

The words are written in these directions

d	z	s	i	e	y	r	r	e	b	w	a	r	t	s
o	g	t	v	s	z	v	m	o	l	h	n	x	u	k
p	k	i	q	j	b	n	n	u	c	l	e	u	s	m
a	y	c	c	d	i	o	n	i	c	d	e	e	s	w
e	h	k	q	f	n	i	a	t	e	a	x	p	j	f
p	t	l	c	o	a	t	n	w	l	b	m	u	l	p
b	l	e	w	e	r	a	s	e	l	e	i	m	b	k
r	a	b	e	t	y	s	c	n	p	o	e	d	v	l
n	t	a	d	u	f	i	r	i	b	m	t	a	c	o
e	e	c	l	s	i	o	a	p	a	h	u	r	y	
l	p	k	f	t	s	i	d	u	c	k	o	x	u	w
l	f	r	o	g	s	t	l	l	a	w	l	l	e	c
o	k	s	d	a	i	r	g	j	a	n	t	h	e	r
p	b	m	o	w	o	e	z	g	m	r	e	p	s	y
y	r	a	v	o	n	f	q	t	f	c	y	a	b	l

1. an animal whose embryo has jelly round it
2. a fruit which isn't a real fruit
3. a real fruit which is good to eat
4. the part of the body where a baby grows
5. a fish which looks after its young
6. a tiny one-celled animal
7. the male sex cell of an animal
8. the male sex cell of a plant
9. an animal embryo's food store
10. something which plant cells have, but animal cells don't
11. the cell's control centre
12. an exploding fruit
13. the joining of a sperm with an egg
14. the human embryo from 8 to 38 weeks
15. an animal which feeds its young on milk
16. the splitting of a cell into two
17. an egg-laying animal
18. the part of a flower which produces pollen
19. the part of a flower which produces ovules
20. attracts insects

Electricity

The Greeks knew about it 2500 years ago.

Benjamin Franklin found out more about it when he flew some kites in 1752.

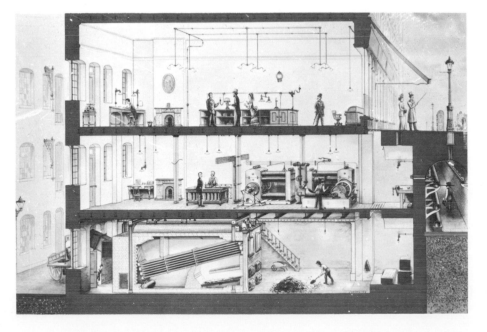

In 1881, a few citizens of London were the first to have it supplied to their homes. Now almost every home in Britain is supplied with it.

What is it?

Electricity, of course. That's what this section is all about.

A polythene rod can do some strange things after being rubbed with a woollen duster. It can pick up dust. It can even make your hair stand on end! These things happen because the rubbing has **charged** the rod.

A rubbed polythene rod attracting hair

Electrical charges

Inside the polythene (and every other substance), there are millions of tiny **electrical charges**. There are two kinds of charge. One is called **positive** (+). The other is called **negative** (−).
Before rubbing, the rod has equal numbers of positive and negative charges. These charges balance each other. The rod is electrically **neutral**. This balance can be upset, however.

Rubbing upsets this balance. When the rod is rubbed with the duster, negative charges travel *from* the duster *to* the rod. The rod now has more negative charges than positive charges. It has become **negatively charged**. Because the negative charges have left the duster, it now has more positive charges than negative. It has become **positively charged**.

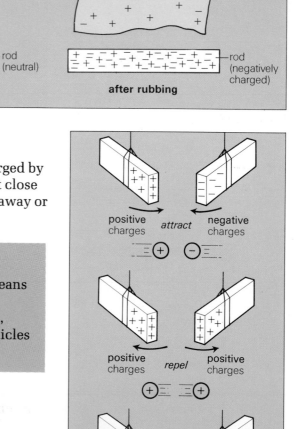

Attraction and repulsion between charged objects

Glass, wood, plastic and many other substances can become charged by rubbing. Charged objects affect each other when they are brought close together. They either pull on, or **attract** each other; or they push away or **repel** each other. What happens depends on the charges:

> **Did you know?**
> - Rubbing gives the polythene rod **static electricity** ('static' means 'not moving').
> - Atoms contain negatively charged particles called **electrons**, positively charged particles called **protons**, and neutral particles called **neutrons**.

1 How many kinds of electric charges are there? What are they called? ▲
2 What happens to an object to make it: a) positively charged   b) negatively charged? ▲
3 Most substances are normally neutral. What does this mean? ▲
4 Two charged polythene rods repel each other. Suggest why.
5 A housewife cleaned a dusty mirror with a dry duster. Next day, it was dustier than ever. What had happened?
6 James was wearing a nylon T-shirt and a woollen jersey. When he pulled off his jersey, it was attracted to his T-shirt. Why?

7.1 Moving charges

You can do some hair raising experiments with a **Van de Graaff generator**! This is a machine for building up and storing electric charges. When it is switched on, negative charges collect on the dome. The dome becomes negatively charged.

Storing charge – who, *me*?
If you stand on a piece of plastic and touch the dome, your body collects extra negative charges from the dome. That's when your hair stands on end. Each hair becomes negatively charged. The hairs repel each other.

Losing charge
If you then touch a water tap, your body loses these charges. Your hair goes back to normal. The charges flow from your body, along the water pipe to the earth. You can feel them go! Your fingers tingle!

Collecting extra charge...   ...and losing it

Flow of charge
There are two good ways of showing this flow of charge. You can connect a neon bulb between the generator and the water tap. The bulb glows as the charges flow through it. You can also connect in a meter which measures small electric currents. The flow of charge through the meter makes the meter needle move. But that's not really surprising, since:

an electric current is a flow of electric charges.

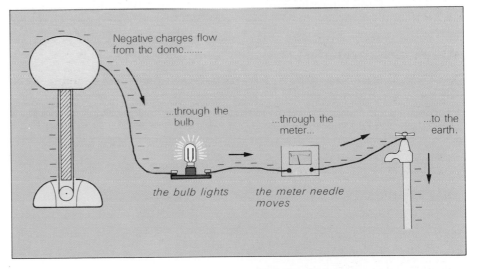

Negative charges flow from the dome.......
...through the bulb
...through the meter...
...to the earth.
the bulb lights   *the meter needle moves*

1 What does a Van de Graaff generator do? ▲
2 a) Why does your hair stand on end when you use the generator to charge yourself up?
   b) Why does your hair go back to normal when you touch a water tap? ▲
3 What is an electric current? ▲
4 Why should you stand on plastic or rubber if you want to charge yourself up?
5 **Try to find out:** about some other machines which use static electricity.

Did you know?

- Metals allow negative charges (**electrons**) to flow through them easily. Plastic and rubber don't.
- Huge Van de Graaff generators have been used by scientists preparing man-made elements.

Things which rub on each other can become charged, and highly charged objects can produce sparks!

The world's biggest, most spectacular sparks are produced by highly charged thunderclouds. All clouds become charged because they contain tiny crystals of ice which are constantly moving and rubbing on each other. A thundercloud can become so highly charged that huge numbers of electrons jump the gap between one cloud and another or between a cloud and the Earth. This avalanche of electrons is a flash of lightning.

Clothes made of man-made fibres like nylon and terylene quickly build up charges when they rub. You can become charged up simply as a result of your clothes rubbing on each other. Sparks are produced when you become so highly charged that electrons jump between you and an earthed metal conductor (like a gas tap or a water tap).

Normally these small sparks do no damage. You may feel a slight shock but no more. But in a factory using flammable materials, even the smallest spark can produce an explosion. That's why workers have to wear special shoes and clothes which don't build up charges. And that's why visitors who wear rubber-soled shoes have to wear special conducting strips. These strips allow any charges which build up to run to earth.

Did you know?

- A flash of lightning can travel at 140 000 km per second.
- A flash of lightning lasts only one millionth of a second.

1 Why do clouds become charged? ▲
2 What is a flash of lightning? When is lightning produced? ▲
3 What special precautions are taken in factories using flammable materials? Why are these precautions taken? ▲
4 Have you ever produced sparks? If so, explain what happened.
5 Aeroplanes become charged up when they fly through clouds. Explain why.
6 **Try to find out:** a) what causes a clap of thunder b) why tall buildings have lightning conductors.

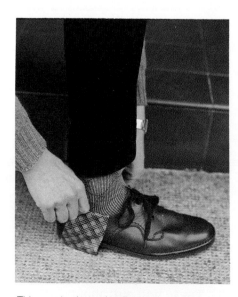

This conducting strip allows any charge to flow to earth

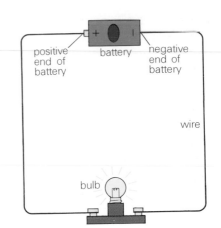

The rest of this section is about electric currents flowing through **electric circuits**:

An electric circuit is a path along which electricity can flow.

You can see an electric circuit in the picture opposite. It is a circuit for lighting up a torch bulb. To make up a circuit like this, you need a battery, a bulb and two wires. The wires are used to join the battery to the bulb. The battery's job is to push an electric current through the wires and the bulb. The bulb lights up as the current flows through it.

Meters and switches

If you want to measure the size of the current, you have to put a meter into the circuit. Current is measured in **ampères** (A) or **amps** for short. The meter for measuring current is called an **ammeter**.

If you want to turn the bulb on and off easily, you should use a **switch**. A switch works by opening and closing a gap in the circuit. When the switch is turned off, a gap opens up. This stops the flow of current all round the circuit. When the switch is turned on, the gap is closed. This makes a **complete circuit** (a circuit with no gaps) and a current can flow.

Here is a picture of a circuit with an ammeter and a switch in it. Beside it is a diagram of the same circuit using electrical symbols:

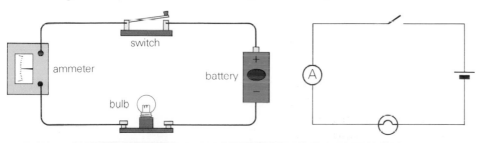

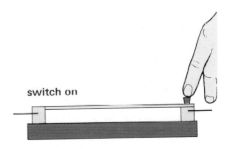

Did you know?

- It is possible to build thousands of tiny electrical circuits called **microcircuits**, on pieces of silicon a few *millimetres* square.
- If a current of 1 amp flows through a bulb, 6 million million million electrons flow through the bulb each second.

1 What is meant by   a) an electric circuit   b) a complete circuit? ▲
2 What does each of the following do in an electric circuit:
   a) battery   b) wires   c) ammeter   d) switch? ▲
3 What is the energy change in: a) the bulb   b) the battery?
4 a) Give one good reason why electricians use symbols, not drawings.
   b) Ⓐ is the symbol for an ammeter. From the circuit diagrams above, work out the symbols for a battery, bulb and switch.
5 What are readings **A**, **B**, **C** and **D** on the ammeter scale shown on the right?
6 **Try to find out:** where microcircuits are used.

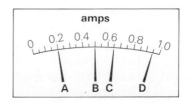

7.2 Conductors and insulators

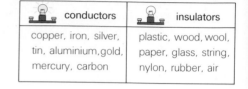

Copper wires are used to connect up electrical circuits. Copper is a **conductor** – a substance which allows electricity to flow through it. You can't use string to join up a circuit! String is an **insulator** – a substance which does not allow electricity to flow through it.

You can use this circuit to test a substance to see whether it is an insulator or a conductor:

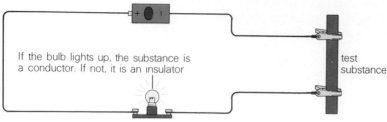

If the bulb lights up, the substance is a conductor. If not, it is an insulator

test substance

💡 conductors	💡 insulators
copper, iron, silver, tin, aluminium, gold, mercury, carbon	plastic, wood, wool, paper, glass, string, nylon, rubber, air

From the table, you should be able to work out that:

metals are conductors
non-metals are insulators (except for carbon which is an important non-metal conductor).

Conductors and insulators both have their uses.

Conductors are used to carry current. Electric current is carried round your home by copper wires.

Insulators are used for safety. The electricity used in your home can be dangerous – you could easily be killed if you touched a bare wire. That is why plugs and wires are covered with insulators like rubber or plastic.

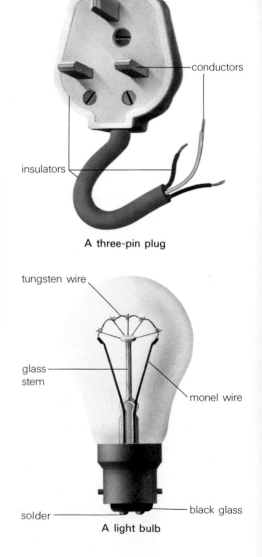

conductors

insulators

A three-pin plug

tungsten wire

glass stem

monel wire

solder

black glass

A light bulb

Did you know?

- A fireman's helmet is made of fibreglass, a non-conductor. Firemen entering burning buildings are often in danger from bare electric wires.
- If the insulation on a cable wears through, you should replace the cable. Never cover bare wires with tape.

1 What is an insulator? What is a conductor? ▲
2 Explain how you would test a substance to see if it was an insulator or a conductor.
3 Why are electric wires: a) made of copper b) coated with plastic? ▲
4 Make separate lists of insulators and conductors from these: *rope, nylon comb, coin, jersey, knife blade, cork, cooking foil.*
5 a) Why do firemen not wear metal hats? ▲
   b) Fishing near electricity wires is always dangerous, but using a carbon fibre rod makes it even more unsafe. Explain why.
6 A light bulb contains insulators and conductors. Which parts are made of insulators and which of conductors?

7.2 Making the right connections

Could you use torch bulbs and batteries to make a quiz board which lights up when you give the right answer? Could you use them to make a 'stop-go' railway signal? The answer is 'yes' – if you think about it. It's all a question of making the right connections.

> **WARNING: Never use mains electricity in experiments like these.**

The quiz board

What you need:
a torch battery, a bulb in its holder, 3 wires, some strips of tinfoil, some sticky tape and a question card.

What you have to do:
1 Work out the correct answers.
2 Work out how you could make the bulb light up when you give each correct answer.
3 Draw a careful diagram of your quiz board to explain how it works.
4 Make the quiz board (in school or at home).

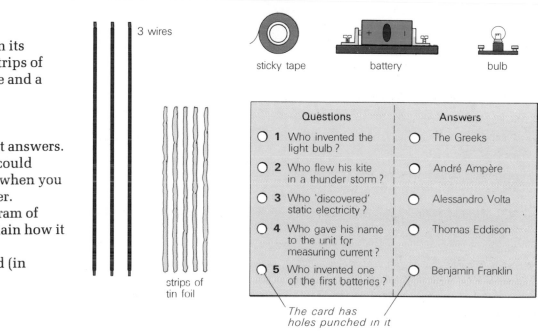

3 wires

sticky tape   battery   bulb

strips of tin foil

Questions	Answers
O 1 Who invented the light bulb?	O The Greeks
O 2 Who flew his kite in a thunder storm?	O André Ampère
O 3 Who 'discovered' static electricity?	O Alessandro Volta
O 4 Who gave his name to the unit for measuring current?	O Thomas Eddison
O 5 Who invented one of the first batteries?	O Benjamin Franklin

The card has holes punched in it

The railway signal

What you need:
two bulbs (one red, one green), a battery, four wires, a 2-way switch and a wooden stand.

What you have to do:
1 Design a circuit which will allow either the green light or the red light to be on. Use one switch to control both lights.
2 Draw a diagram of your circuit and explain how it works.

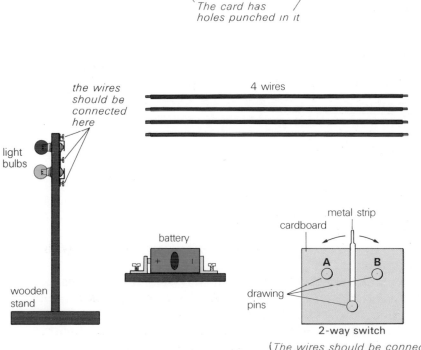

the wires should be connected here

4 wires

light bulbs

battery

wooden stand

metal strip

cardboard

A   B

drawing pins

2-way switch

(*The wires should be connected to the drawing pins. The metal strip can move to make contact with either **A** or **B***)

7.3 Pushing the current

When a battery is working, the energy change is:

chemical energy ⟶ **electrical energy.**

The battery's chemical energy is used up pushing a current round the circuit.

The 'electrical push' which the battery gives to the current is called the **voltage**. It is measured in **volts (V)** on a **volt-meter**.
Different batteries produce different voltages. The bigger the voltage supplied by the battery, the bigger will be the current that flows in the circuit:

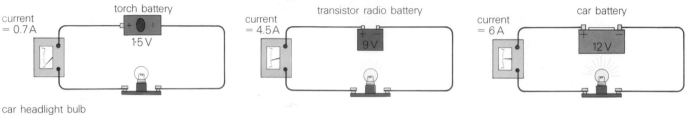

current = 0.7 A · torch battery · 1·5 V

current = 4.5 A · transistor radio battery · 9 V

current = 6 A · car battery · 12 V

car headlight bulb
(it works properly when a current of 6 A flows through it)

A torch battery only produces 1.5 V, but you can produce a bigger voltage by joining up several batteries. The batteries have to be joined in **series** (one after the other). They must also be lined up in the same direction so that they push together.

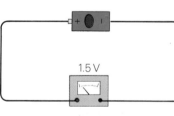

1.5 V

voltmeter

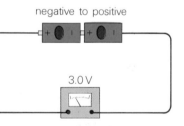

negative to positive

3.0 V

4.5 V

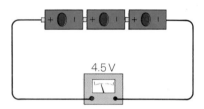

0 V

Did you know?

- The voltage of the electricity mains is 250 V.
- An electric eel has muscle cells which act like tiny batteries. It gives its prey a 600 V electric shock!

1 What is meant by: a) a voltage b) joining torch batteries in series? ▲
2 What happens to the size of the current when the voltage increases? ▲
3 a) The headlight bulb does not even glow when it is connected to the torch battery. Why is this?
   b) How could you use torch batteries to make the bulb light?
4 One 'electric' muscle cell in an electric eel produces 0.1 V. How many 'electric' muscle cells does an eel have?
5 Explain the voltage readings in the circuit diagram shown on the right.

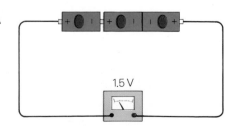

1.5 V

7.3 Portable power packs

Going further

If you are going to be absolutely correct, you should call a torch battery a **dry cell**. It is called *dry* because it has no liquid in it. A **battery** is made up from a collection of cells joined together. A 9 V transistor radio battery contains 6 small cells joined in series. A 12 V car battery contains six 2 V lead-acid cells joined in series.

There are many different cells in use today. Here are six of them, with their advantages and disadvantages:

The dry cell is the most commonly used cell. It is often used in torches – it is cheap, easily carried and has no liquid to spill.

It cannot give big currents and its voltage falls during use. It has to be thrown away once it has gone flat.

The mercury cell can be made into small 'button cells'. Even very small mercury cells can produce large currents for a short time, or small currents for a long time.

Mercury cells are expensive.

The lead-acid cell can produce large currents. A battery of lead acid cells can produce the large current needed to start a car. The cell can be recharged when flat.

It is heavy and contains acid which can spill.

The reserve cell does not work until salt water is added. Then it can give a high current for a short time or a small current for a long time.

The nickel-cadmium cell is the 'rechargeable battery' sold in the shops. It is light and is completely sealed.

It cannot give very large currents and is expensive.

The lithium cell is small and light. It is very reliable and long lasting.

Lithium cells are expensive.

1 What is a battery? How is a car battery made up? ▲
2 Hearing aids only use small currents. Suggest two reasons why a mercury cell is a good cell for running a hearing aid. ▲
3 a) Suggest reasons why the dry cell is the most commonly used cell.
    b) If you leave a torch on, the bulb gradually gets dimmer. Why? ▲
4 One of the six cells can be found in a life raft. Which cell is this? What might it be used for?
5 **Try to find out:** what heart pacemakers are. Then suggest why lithium cells are used to power them.

Did you know?

- Lead-acid batteries have been used to run all-electric cars. But the batteries are very heavy. The cars can't travel very far before the batteries have to be recharged.
- A new, lighter sodium-sulphur cell may soon be able to power all electric cars.

7.4 Resistance

You can use any kind of metal wire to connect up an electrical circuit. But the size of the current depends on the metal you use:

It is more difficult for the battery to push a current through nichrome than through copper. That's why a smaller current flows in the circuit with the nichrome wire connectors.

Electrical resistance

Nichrome **resists** the current more than copper. Nichrome has a bigger **resistance** than copper.

Every material has an electrical resistance. The greater the material's resistance, the smaller is the current which flows through it. Conductors like copper and aluminium have very low resistances. They carry large currents well. But insulators have very large resistances. They only allow very small currents to flow through them.

When a current flows through a wire, electrical energy is changed to heat energy.

When a current is pushed through a high resistance wire by a large voltage, large amounts of heat are produced. That is why wires made of nichrome are used to make the heating elements in electric fires and kettles.

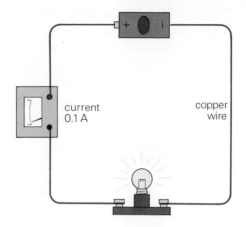

current 0.1 A
copper wire

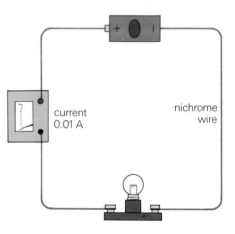

current 0.01 A
nichrome wire

Did you know?

- Glass normally has a resistance one million, million, million times greater than copper.
- The **ohm**, the unit of resistance, was named after a German schoolteacher George Ohm. Around 1826 he lost his job for doing experiments on resistance. Experiments were frowned on!

---

1 Copper has a smaller resistance than nichrome. What does this mean?
2 What can you say about the resistance of   a) a conductor   b) an insulator? ▲
3 a) Why is nichrome used in the heating element of an electric kettle? ▲
   b) Why does a T.V. heat up when you use it?
4 The resistance of a car headlight bulb changes after it has been switched on. The resistance increases as the bulb heats up. How will this affect the current flowing through the bulb? Explain.
5 **Try to find out:** what a resistor is, and how to tell its resistance from the colour code.
6 The diagram on the right shows how the 'rings' on an electric cooker are made. Explain why:
   a) the heating wire must have a high resistance
   b) the packing must have a very high resistance.

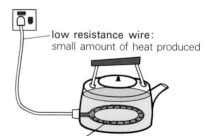

low resistance wire:
small amount of heat produced

high resistance wire:
(inside the kettle's element)
A lot of heat is produced here

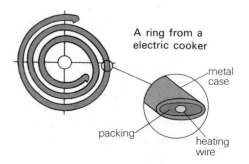

A ring from a
electric cooker

metal case

packing

heating wire

7.4 The resistance of a wire

The resistance of a wire depends on:

	what it's made of	how long it is	how thick it is
low resistance	copper	short	thick
high resistance	nichrome	long	thin

A short, thick copper wire will have a low resistance. A long, thin nichrome wire will have a high resistance.

Some electrical work needs high resistance wire. Most electrical work, however, needs low resistance wire. You must choose the correct wire for the job.

Wires which are used to carry current should have a low resistance. When low resistance wires are used, only a small amount of electrical energy is changed to heat energy as the current flows. High resistance wire is used in electrical heaters. This kind of wire produces large amounts of heat when electricity is pushed through it.

If you want to make gradual changes in the current flowing in a circuit, you can use a **variable resistance** ('variable' means 'changing'). This is really just a resistance wire with a sliding contact. By moving the contact, you can change the length of resistance wire through which the current flows.

When the contact is moved to the right, there is more of the resistance wire in the circuit. This gives the circuit a higher resistance. It is harder for the current to flow – the current falls, and the bulb becomes dimmer.

Electricity cables are made of thick aluminium wires

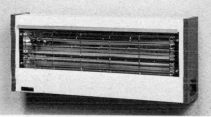
The heating element of an electric fire is a long coil of thin nichrome wire

This is a variable resistance. The sliding contact moves along the top bar

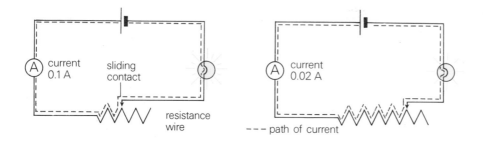

1 Which has the bigger resistance, a short thick copper wire or a long thin nichrome wire? Explain your answer. ▲
2 a) Why is thick copper wire good for carrying current?
　b) Why are long thin coils of nichrome wire used in an electric fire?
3 What is a variable resistance used for? How does it work? ▲
4 A radio's volume control has a variable resistance. Explain what happens when you turn down the volume.
5 **Try to find out:** why the cable to a cooker is thicker than the cable to a table light.

Did you know?

- Carrying electricity by overhead cables is efficient. Only 3% of the energy is wasted as heat.
- The controls of model electric cars and trains have variable resistances.

7.4 Is anything really an insulator?

It's useful to remember that 'metals are conductors, non-metals are insulators'. But, strictly speaking, this is not the whole story.

All substances conduct to some extent. Most non-metals, however, have such high resistances that they only carry very small currents. In fact, the currents carried by non-metals are normally so low that they can be ignored. That's why non-metals are given the name 'insulators'.

substance	conductivity
Silver	66 000 000
Copper	64 000 000
Aluminium	44 000 000
Steel	5 500 000
Water	0.0001
PVC	0.000 000 01
Air	0.000 000 000 000 025
Rubber	0.000 000 000 000 01

You can tell how well a substance conducts from its **conductivity**. A table of conductivities is given at the top of the page. The bigger the conductivity, the better the conductor. The conductivity units have been left out of the table to make it simpler.

If the voltage is high enough, even the poorest conductor will carry a small current.

For air, the voltage needed is very high. Normally, air is a very good insulator, but at very high voltages, electrons can flow through it. 33 000 V will make electrons jump across a 1 cm air gap.

For water, another poor conductor, the voltage needed is not nearly so big. Even at mains voltage (250 V), water can carry a current big enough to kill a human. That's why you have to be very careful to keep water away from electricity.

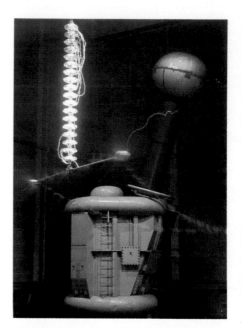

Testing an electricity pylon insulator. This one started conducting at 880 000 V

Did you know?

- Wet skin conducts better than dry skin.
- Metals conduct best at low temperatures. Around $-270\,°C$, some metals have no resistance at all!

1 Why are non-metals: a) not *really* insulators b) called insulators? ▲
2 Why must water be kept away from electricity? ▲
3 Why is the best conductor not used for household wiring?
4 Why is it dangerous to touch electrical switches with wet hands?
5 The insulator being tested in the photograph is used for carrying high voltage electricity. It is designed so that the whole of its surface never gets wet. Why is this?
6 **Try to find out:** about special precautions which must be taken when electrical fittings are put in bathrooms.
7 The illustration on the right shows a 'lie detector'. When the 'victim' lies, her skin resistance goes down and the current goes up. Why?

There arc two ways in which you can connect up bulbs in a circuit. The bulbs can either be connected **in series** or **in parallel**. The way in which you connect them makes quite a difference.

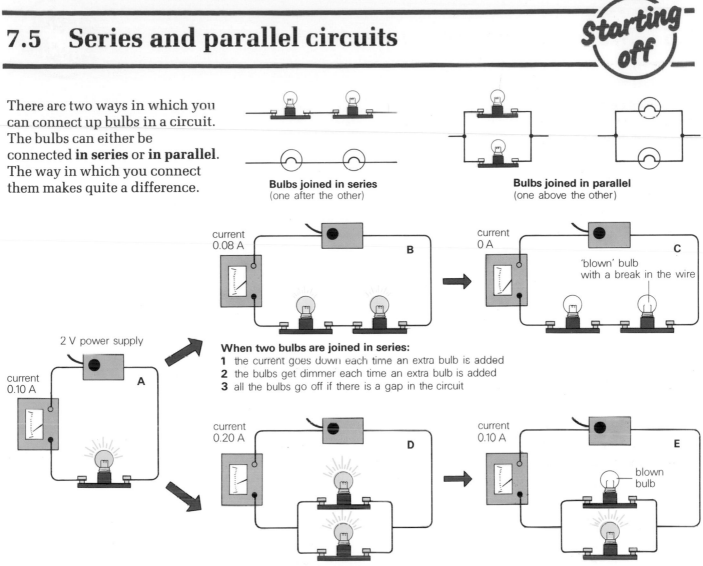

Bulbs joined in series
(one after the other)

Bulbs joined in parallel
(one above the other)

2 V power supply

current
0.10 A

A

current
0.08 A

B

current
0 A

C

'blown' bulb
with a break in the wire

When two bulbs are joined in series:
1   the current goes down each time an extra bulb is added
2   the bulbs get dimmer each time an extra bulb is added
3   all the bulbs go off if there is a gap in the circuit

current
0.20 A

D

current
0.10 A

E

blown
bulb

When two bulbs are joined in parallel:
1   extra current is drawn from the power supply each time a bulb is added
2   the bulbs stay bright even when extra bulbs are added
3   each bulb can go on and off without affecting the others

Did you know?

• The lights in your house are wired in parallel.
• Some Christmas tree lights are wired in series. If one goes out, the others go out, too.

1   Why do the two bulbs in circuit **B** make less light than the one bulb in circuit **A**? ▲
2   Why does no current flow in circuit **C**? ▲
3   Using symbols, draw a circuit with three bulbs in series. Then draw a circuit with three bulbs in parallel.
4   Suggest two reasons why it is better to have house lights wired in parallel and not in series.
5   Are car headlights wired in series or in parallel? Give a reason for your choice.
6   Look at circuit **F**. Which switches will have to be closed to light:
a) bulb **1** only   b) bulbs **1** and **2** together?

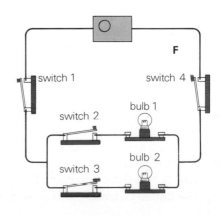

F

switch 1   switch 4

switch 2

bulb 1

switch 3   bulb 2

7.5   Measuring some currents

In a series circuit the same current flows through each part of the circuit. (That's what you might expect. There is only one path for the current to flow round.) Circuits **A** and **B** are **series circuits**.

In a parallel circuit the current divides when it comes to a junction. Part of the current flows through one branch. Part of it flows through the other. The bulbs in circuits **C** and **D** are joined **in parallel**.
When the two branches have the same resistance, the same current flows through each branch. (That's what happens in circuit **C**. Bulbs **1** and **2** are exactly the same.)

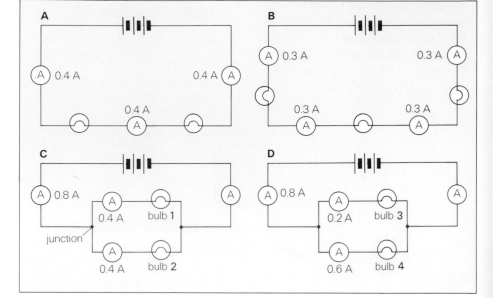

When the two branches have *different* resistances, a bigger current flows through the branch with the *smaller* resistance – it's easier this way! This happens in circuit **D**. Bulb **3** has a bigger resistance than bulb **4**. That's why more of the current flows through bulb **4**. Bulb **4** will glow more brightly because of the bigger current.

In all parallel circuits, one thing is true:

$$\text{The sum of the currents flowing in the branches} = \text{Current flowing in the rest of the circuit}$$

Did you know?

- In the **short circuit** shown in circuit **E** on the right, practically all the current flows through the copper wire because it has a much lower resistance than the bulb.
- Because the resistance is low, the current which flows through a short circuit is very much larger than normal. The wires can get very hot. Short circuits, caused by bare wires touching, often cause fires.

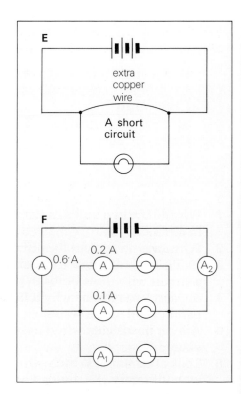

1   Why does the same current flow through every part of a series circuit? ▲
2   What happens to the current at a junction in a parallel circuit? ▲
3   Why does the same current flow through bulbs **1** and **2** in circuit **C**? ▲
4   Why does more current flow through bulb **4** in circuit **D**? ▲
5   What are the currents flowing through meters A_1 and A_2 in circuit **F**? Explain your answers. Which bulb glows brightest?
6   Circuit **E** was set up first without the extra copper wire. The bulb lit up. When the extra copper wire was added, the bulb went out and the battery quickly went flat. Explain why.

7.5 Carrying the current round your home

There is one big difference between the electricity which is supplied by a battery and the 'mains electricity' which is supplied to your home. The electricity supplied by a battery is **direct current** or **d.c.** It flows round the circuit in one direction only.

Mains electricity is **alternating current** or **a.c.** The current *changes direction* 50 times every second!

Household circuits

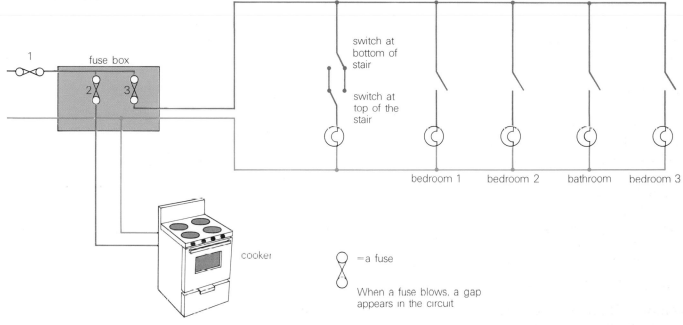

= a fuse

When a fuse blows, a gap appears in the circuit

Electricity is carried to and from your home by two wires, the **live** wire and the **neutral**. At the fuse box, these wires are connected to a number of circuits. Each circuit has a live wire and a neutral wire.

The diagram above shows you two of the circuits you would find in a two-storey house. One circuit is for upstairs lights. The other circuit is for the cooker. In the house, there are likely to be four other circuits (downstairs lights, upstairs power sockets, downstairs power sockets and immersion heater).

A domestic fuse box

1 What do 'a.c.' and 'd.c.' stand for? How are they different? ▲
2 List the circuits which might be found in a two-storey house. ▲
3 Now look at the diagram above:
   a) How are the bulbs connected in the house?
   b) Will the bathroom light be affected by switching on the others? Explain.
   c) Why does the stair light have special switches? How do they work?
4 Why is the cooker on a separate circuit?
5 What will happen if: a) fuse 1   b) fuse 2   c) fuse 3   blows?
6 **Try to find out:** when the off-peak circuit supplies electricity, and why it is supplied at a cheaper rate.

Did you know?

- Power circuits are designed to carry bigger currents than lighting circuits.
- Many houses have an 'off-peak' circuit. It only supplies electricity at certain times of the day.

Electrical plugs

Why does the cable of a television set have a plug at the end of it? Because pushing the plug into a wall socket is the easiest and safest way of connecting the television to the electrical supply.

It is important that the plug is wired up properly. The cable usually has three wires. These are the **live** which is brown, the **neutral** (blue) and the **earth** (green and yellow stripes). The live and neutral wires carry the current. The earth wire is for safety.

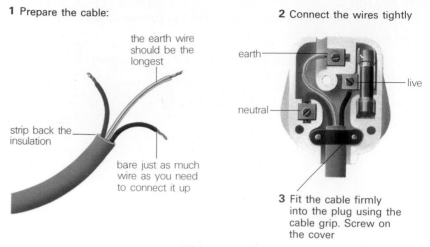

1 Prepare the cable:

the earth wire should be the longest

strip back the insulation

bare just as much wire as you need to connect it up

2 Connect the wires tightly

earth

live

neutral

3 Fit the cable firmly into the plug using the cable grip. Screw on the cover

Wiring a plug

Fuses

Fuses are used for safety, too.
A fuse prevents a circuit from carrying too large a current. In this way, it can cut down the risk of fire. It can also prevent an electrical appliance from being damaged.

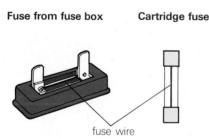

Fuse from fuse box   Cartridge fuse

fuse wire

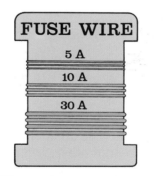

FUSE WIRE

5 A

10 A

30 A

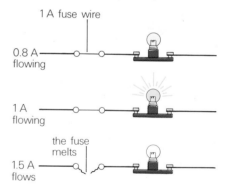

1 A fuse wire

0.8 A flowing

1 A flowing

the fuse melts

1.5 A flows

1 The working part of every fuse is the fuse wire. It is made of a metal with a low melting point. The wire carries the current across the fuse.

2 Different fuse wires can carry different currents. A 5 A fuse wire can carry 5 A but will melt if a bigger current flows.

3 As the current increases, the fuse wire heats up. If the fuse wire becomes too hot, it will melt. This stops the flow of electricity.

1   Make a list of the three wires in a cable, with their colours. ▲
2   Write down what you would do to wire a plug. ▲
3   What is a fuse for? How does it work? ▲
4   The 1 A fuse wire has protected the bulb in the picture above. How? (Clue: What would a big current do to the bulb if there was no fuse?)
5   Why would it be silly to use: a) a 1 A fuse in the plug of a kettle using 3 A   b) a 13 A fuse in the plug of a T.V. using 1.5 A?
6   **Try to find out:** what the old wire colours were, and why they were changed.

Did you know?

- 2 pin plugs have no earth wire connection.
- Fuse wire is a mixture of tin and lead.
- 'Fuse' is the scientific word for 'melt'.

No bad habits, please

'Playing' with electricity can be deadly. Carelessness can be highly dangerous, too. Despite this, many people get into bad habits when they use electricity.

You can see some of the bad habits in the diagrams below. Write a sentence about each one, saying how you think it is dangerous. Then, if you can, write another sentence to say what should be done.

The earth wire

An electrical appliance in good working order, with all wires properly insulated and all connections secure, is safe to use. But worn-away insulation and loose wires can make an appliance dangerous. A faulty appliance can kill you!

It is the **live wire**, carrying high energy electricity, which can cause the problem. If a bare, live wire touches any metal part of an appliance, that part will become live, too. If you are unlucky enough to touch a live appliance, you will be given a serious electric shock. The current runs from the live wire through you to the earth. But if the appliance is properly earthed and fused, an accident like this should never happen.

The reading lamp shown on the right is properly earthed. When it is plugged in, its base (and all other metal parts) are connected to the earth by the earth wire.

If the insulation on the live wire wears through, the lamp can become live. But it won't stay live for long because:

1   the current flows to earth. It flows from the live wire, through the metal case of the lamp and along the earth wire. This produces a short circuit and:

2   the current blows the fuse. This happens because the current in the short circuit is much bigger than the normal current.

3   the electricity supply to the light is cut off. This happens when the fuse blows.

All of this happens in a fraction of a second often with a bang!

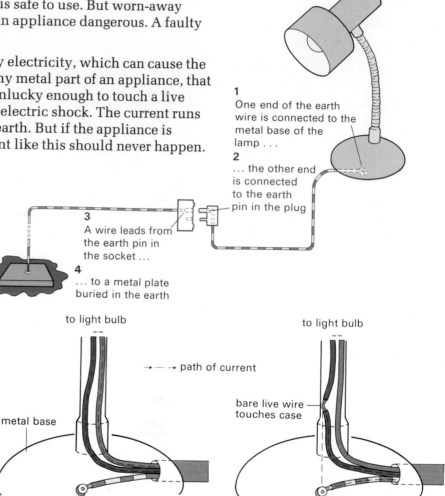

1
One end of the earth wire is connected to the metal base of the lamp . . .

2
. . . the other end is connected to the earth pin in the plug

3
A wire leads from the earth pin in the socket . . .

4
. . . to a metal plate buried in the earth

to light bulb

→ — → path of current

metal base

The current normally flows to and from the light bulb along the live and neutral wire

to light bulb

bare live wire touches case

If the insulation wears through, the current flows through the metal to the earth wire

1   Why can loose connections and broken insulation make an appliance dangerous? ▲
2   Why is a live appliance dangerous? ▲
3   Explain how the earth wire and the fuse prevent an accident from happening if an appliance becomes live. ▲
4   Many parts of electrical appliances are made of plastic. Why does this increase the appliance's safety?
5   **Try to find out:** why the use of plastic water pipes may cause problems with house earthing systems.

Did you know?

● A 250 V electric shock can be fatal. In 1967, however, an American called Brian Latasa survived a 230 000 V shock.

● Appliances marked 'double insulated' have no earth wire. All metal parts of these appliances are surrounded by insulating plastic.

The gases of the air

Round about the Earth is the **atmosphere**. It's a layer of gases, about 15 km thick.

The atmosphere doesn't seem to be very big, certainly not when compared with the size of the Earth. (If the Earth was a globe the size of a football, the atmosphere would be thinner than the coat of paint round the outside!) But the whole atmosphere weighs about 5 000 000 000 000 000 tonnes. And that's a lot of gas!

In the atmosphere, there are several different gases mixed together. This mixture of gases is called **air**.

Air is needed for breathing.

Air is needed for burning.

Air can be useful in other ways too.

You won't need anyone to tell you how important air is. You can't live without it! And so this section is mostly about air. It's about:
- the gases which make up the air
- burning things in air
- why air is so important to life.

8.1 The gases in the air

The main gases mixed together in the air are **nitrogen**, **oxygen**, **argon** and **carbon dioxide**. There are a few other gases, in tiny amounts.

Nitrogen makes up nearly ⅘ of the air.

Nitrogen gas is made up of molecules. Each molecule contains two nitrogen atoms

78% of the air is nitrogen

Nitrogen doesn't burn. In fact, it isn't a very exciting gas. It isn't easy to make nitrogen join up with other chemicals. It puts out most things which are burning.

Oxygen makes up about ⅕ of the air.

An oxygen molecule contains two oxygen atoms

21% of the air is oxygen

Oxygen doesn't 'burn'. When other substances 'burn', they are joining up with oxygen. If a red glowing wooden splint is put in oxygen, the splint burns with a bright flame. This test is used to pick out oxygen from other gases.

A gas jar of air (more than 10 different gases mixed together)

Carbon dioxide makes up only a small part of the air.

0 03% of the air is carbon dioxide

A carbon dioxide molecule contains one carbon atom and two oxygen atoms

Carbon dioxide doesn't burn, and it puts out most things which are burning. But when you shake it up with lime water the clear liquid goes milky white. The other gases don't change the lime water.

Argon makes up only a small part of the air.

1% of the air is argon

Argon gas is made up of single atoms

Argon is one of the **noble gases**. It doesn't normally join up with any other chemicals. It doesn't burn. It puts out things which are burning.

Did you know?

● On Mars and Venus, the atmospheres are mostly made up of carbon dioxide. On Saturn, the atmosphere contains hydrogen, helium and methane (the same chemical as North Sea Gas). Mercury and the Moon have almost no atmosphere at all.

1 About oxygen, nitrogen, carbon dioxide and argon:
  a) Put the gases in order, most plentiful first.
  b) Which gases put out things which are burning? ▲
  c) How much of each gas is there in 100 cm³ air?
  d) Which gases are elements, and why? (see page 44)
  e) In what way is argon the 'odd man out'?
2 How can you pick out: a) oxygen b) carbon dioxide from other gases? ▲
3 **Try to find out:** the names of other gases in the air.

There is only a small amount of water vapour in the air. If there is too much, it turns to liquid !

8.1 Using the gases

The gases in the air can be used to do a variety of jobs. That's why the gases are separated from each other and stored in cylinders. The cylinders have colour codes which tell you which gas is inside. Here are just a few of the uses:

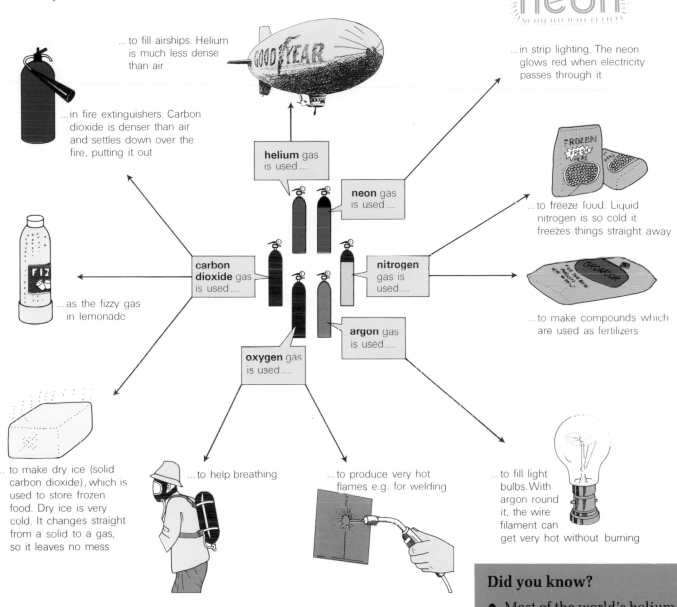

...in fire extinguishers. Carbon dioxide is denser than air and settles down over the fire, putting it out

...to fill airships. Helium is much less dense than air

helium gas is used.....

neon gas is used.....

...in strip lighting. The neon glows red when electricity passes through it

...to freeze food. Liquid nitrogen is so cold it freezes things straight away

...as the fizzy gas in lemonade

carbon dioxide gas is used.....

nitrogen gas is used.....

...to make compounds which are used as fertilizers

oxygen gas is used.....

argon gas is used.....

... to make dry ice (solid carbon dioxide), which is used to store frozen food. Dry ice is very cold. It changes straight from a solid to a gas, so it leaves no mess

...to help breathing

...to produce very hot flames e.g. for welding

...to fill light bulbs. With argon round it, the wire filament can get very hot without burning

1 Write down one use for: a) nitrogen b) carbon dioxide c) neon. ▲

2 Why are: a) airships filled with helium b) light bulbs filled with argon? ▲

3 Explain why: a) dry ice is better than ordinary ice b) carbon dioxide is a good gas for using in fire extinguishers. ▲

4 Why would you expect to find cylinders of oxygen in: a) hospitals b) blacksmiths' workshops? Where else might you find them?

5 **Try to find out:** why deep sea divers are not supplied with ordinary air.

Did you know?

● Most of the world's helium comes from underground gas fields in the U.S.A. There it is found mixed with fuel gases.

● Divers who work at depths below 75 m breathe oxygen mixed with helium. This makes their voices sound squeaky!

Getting pure oxygen and nitrogen from the air may seem to be an impossible task. In air, all the different gases are completely mixed through each other.

Separating out pure oxygen and nitrogen is very difficult – as long as they stay as gases. But separation can be carried out much more easily by changing the gases to liquids because each liquid has a different **boiling point (b. pt.)**.

To produce pure samples of oxygen and nitrogen, the air is gradually cooled. As the temperature falls, the water vapour freezes and is removed. Carbon dioxide is also removed. The rest of the air is then cooled down a long way and the pressure is increased. This causes most of the gases to liquefy, producing **liquid air** at a temperature of about $-200\,°C$. The liquid air is allowed to warm slowly until the nitrogen (b. pt. $-196\,°C$) boils off as a gas. The oxygen (b. pt. $-183\,°C$) is left behind as a liquid.

The oxygen and nitrogen have been separated by **fractional distillation**. Fractional distillation is used to separate liquids with different boiling points. When a mixture of liquids is heated up, the liquid with the *lowest* boiling point boils off *first*.

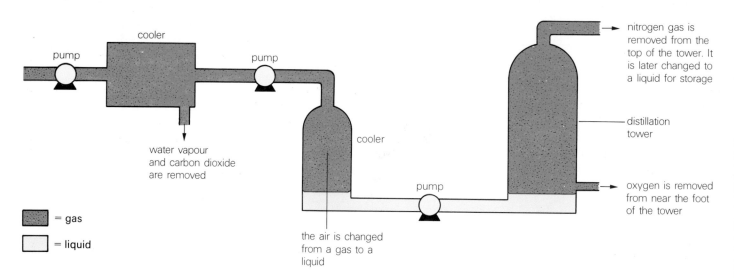

This diagram shows a few of the steps in the fractional distillation of liquid air. The real process is much more complicated

1 What happens to a liquid in distillation? (see page 62)
2 What is fractional distillation used for? ▲
3 a) How is liquid air produced?
  b) How are pure oxygen and pure nitrogen produced from liquid air? ▲
4 Argon has b. pt. $-186\,°C$. Neon has b. pt. $-246\,°C$.
  a) Put argon, neon, oxygen and nitrogen in order, with the lowest boiling point substance first.
  b) How would you separate argon from liquid air?
5 **Try to find out:** what happens in a distillery.

Did you know?

- Space rockets carry liquid oxygen so that they can burn fuel in space.
- Large rockets carry more than 2 million litres of liquid oxygen.

8.2 What happens when things burn?

Look what happens in these three experiments on burning:

1 A candle burns brightly in air. But if its air supply is cut off, it goes out.

About ⅕ of the air is used up. Then the candle goes out. This is because **the candle uses the oxygen in the air as it burns**.

2 Hot carbon burns in oxygen. Carbon dioxide gas is produced.

Carbon dioxide contains carbon and oxygen atoms joined together. This experiment shows that **when carbon burns, it joins up with oxygen**.

3 Magnesium ribbon burns with a brilliant white flame. A white ash called magnesium oxide is produced.

If the magnesium is burned in a crucible, the crucible and contents are found to weigh more at the end of the experiment than they did at the beginning. This is because **the burning magnesium joins up with oxygen from the air**.

the test tube cuts off the candle's air supply

the water rises to take the place of the oxygen used

carbon

oxygen

lime water

lime water turns milky because carbon dioxide is present

the lid has to be lifted to let air in

crucible
2·4 g magnesium

4·0 g magnesium oxide

These three experiments make the point that:

oxygen is needed to make things burn.

A burning element joins up with oxygen. A new compound – the element's **oxide** is produced. Energy (mostly heat) is also given out.

1 a) Why does the water rise in the test tube in experiment 1? ▲
    b) Why does the lime water go milky in experiment 2?
    c) Why does the crucible get heavier in experiment 3? ▲
2 What is used up and what is produced when an element burns? ▲
3 a) Why is magnesium used in some photographic flash bulbs?
    b) Why must flash bulbs contain oxygen?
4 Petrol is a compound made up of two elements. When petrol burns, carbon dioxide and water (hydrogen oxide) are produced. Which elements are in petrol?

Did you know?

● Magnesium burns with such a bright light that fine threads of magnesium are used in some photographic flash bulbs.
● Midget flash bulbs contain a metal called zirconium. It burns even brighter than magnesium.

8.2 Putting out fires

This is the 'fire triangle'. Its three sides are heat, fuel and air (or oxygen). These are the three things which are needed for burning.

A triangle has to have three sides. If you take away one side, it will collapse. Heat, fuel and air are all needed for burning. If one of them is missing, there won't be a fire.

Now you have to use the fire triangle to help you to write some sentences about putting out fires.
Write each sentence by matching up two boxes, one from each side:

To put out a chip pan fire, you should...

To put out a burning pool of petrol, you can...

To put out a bonfire, you should...

To stop a forest fire from spreading, foresters...

If someone's clothes catch fire, you should...

When a fire starts in one part of a house...

... roll her (or him !) in a rug or curtain. (This cuts off the air supply)

... close up windows and doors to stop draughts (and shut off the air supply)

... hose it down with water. (This cools down the fire)

... cover it with the lid or a damp cloth. (This shuts off the air supply)

... cut down some of the wood. (This takes away the fire's fuel)

... throw a bucket of sand over it. (This shuts off the air supply)

WARNING
- You should never use water to put out petrol fires, electrical fires or chip pan fires.
- Try to work out, or find out, why not.

8.2 Burning problems

The energy from burning fuels is really important for our way of life. It is used for heating homes, for cooking food and for driving cars. It is used in power stations to make electricity and in industry for different kinds of heating.

But there are snags in burning a lot of fuel. Burning fuels cause **air pollution**.

Burning fuels make the air dirty

Hundreds of thousands of tonnes of sooty smoke pour out of British chimneys each year. The soot is made up of tiny specks of carbon. It is produced when fuels don't have enough air to burn completely. Smoky chimneys make cities dirty. Smoky cities are unhealthy, too. That's why Governments and City Councils have taken steps to cut down the amount of soot in the air. Governments have passed 'Clean Air Acts' which have made factories clean up the smoke they produce. These Acts have also allowed City Councils to set up **smokeless zones** – areas of the city where no smoky fuels can be burned.

Sulphur dioxide is released into the air at petrol refineries

Burning fuels produce harmful gases

One of these harmful gases is **sulphur dioxide**. Some fuels, like coal and oil, contain sulphur. The sulphur dioxide is produced when these fuels burn.

Sulphur dioxide is harmful because it dissolves in water to make an acid. The acid produced when sulphur dioxide dissolves in rain, eats away stone and metal, and poisons plants. Sulphur dioxide can damage your lungs if you breathe a lot of it.

Carbon monoxide is another harmful gas. Like soot, it is produced when fuels don't have enough air to burn completely. Badly adjusted motor cars and gas fires produce lots of carbon monoxide.

Carbon monoxide is very poisonous. It is particularly dangerous because it has no smell. Faulty gas fires and leaky car exhausts have caused many fatal accidents.

Nelson's column, London: before and after cleaning

Did you know?

- There are more than 200 million cars in the world. Each can produce 1000 kg of harmful gas each year.
- In Tokyo, traffic policemen wear gas masks.

1 What is: a) soot b) a smokeless zone? ▲
2 Why can burning a fuel release sulphur dioxide into the air? Why is this harmful? ▲
3 A yellow bunsen flame is sooty. What does this tell you about the burning gas? How could you make the flame burn cleanly?
4 Why must you never run a car in a small, closed garage?
5 **Try to find out:** why the burning of fuels in Britain causes problems in Sweden.

These trees were killed by small amounts of sulphur dioxide in the air

Breathing and producing energy go 'hand in hand'. Your body can't do one without doing the other.

Your body gets its energy from a chemical reaction which goes on in its living cells. The reaction is a bit like a slow 'burning up' of food. Food and oxygen join up. Energy (mainly heat and movement) is produced. Carbon dioxide and water are made, too.

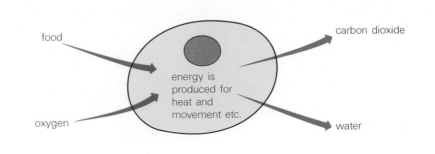

What goes on in a cell

This means that, when they are producing energy, your cells:

- use up oxygen
- produce carbon dioxide and water.

And so, to keep producing energy, your body must:

- be supplied with oxygen
- get rid of carbon dioxide (which could poison it)
- get rid of some of the water.

That's where breathing comes in. You breathe in fresh air. It contains the oxygen you need. You breathe out stale air, with carbon doxide and water vapour in it.

Did you know?

- This 'burning up' of food to give energy is called **respiration**. All living things carry out respiration.
- Your cells don't go on fire! The energy is released slowly. Chemicals called **enzymes** control this slow release of energy.

If you breathe on a cold window, you can see some of the water from your breath

1 a) What does a cell need to produce energy?
   b) How is energy produced in the cell? ▲
2 What is meant by respiration? ▲
3 Why do you have to breathe to produce energy? ▲
4 Why does a cold window 'steam up' when you breathe on it?
5 How does the air in your classroom change during a lesson?
6 **Try to find out:** what is meant by 'ventilation', and why it is important.
7 Which of the jars of air illustrated on the right will:
   a) let a splint burn longer?
   b) turn lime water milky?
   Explain your answers.

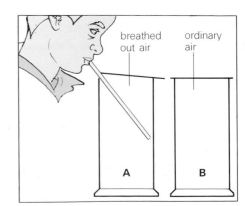

8.3 Breathing [2]

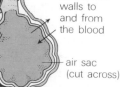

Every living cell in your body:

- has to get a supply of oxygen
- has to get rid of carbon dioxide and water.

So how do these gases get to and from the cells?

The answer is in two parts. First, the gases have to get into and out of your body. This really happens in your **lungs**. Then the gases have to be carried from the lungs to the cells, and then back to the lungs. This job is done by your **blood**.

When you breathe in, you take fresh air into the lungs.

The fresh air travels:

Your body gets rid of carbon dioxide and water vapour when you breathe out.

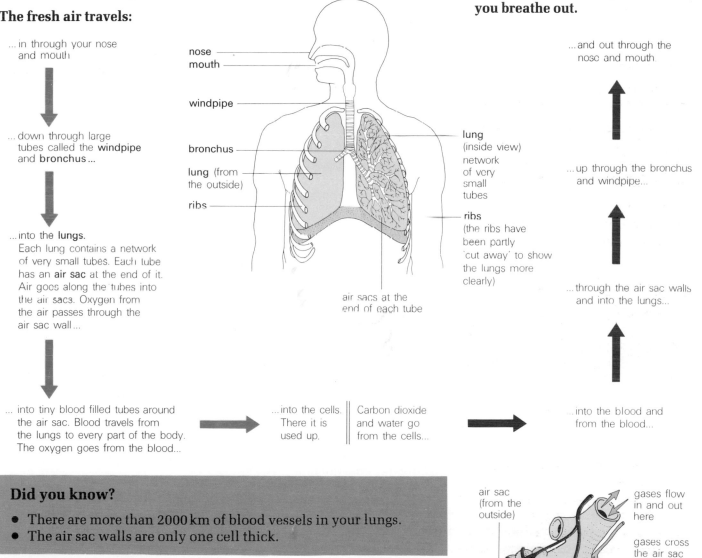

...in through your nose and mouth

...down through large tubes called the **windpipe** and **bronchus**...

...into the **lungs**.
Each lung contains a network of very small tubes. Each tube has an **air sac** at the end of it. Air goes along the tubes into the air sacs. Oxygen from the air passes through the air sac wall...

... into tiny blood filled tubes around the air sac. Blood travels from the lungs to every part of the body. The oxygen goes from the blood...

...into the cells. There it is used up.

Carbon dioxide and water go from the cells...

...and out through the nose and mouth.

...up through the bronchus and windpipe...

...through the air sac walls and into the lungs...

...into the blood and from the blood...

nose
mouth

windpipe

bronchus

lung (from the outside)

ribs

lung (inside view) network of very small tubes

ribs (the ribs have been partly 'cut away' to show the lungs more clearly)

air sacs at the end of each tube

air sac (from the outside)

gases flow in and out here

gases cross the air sac walls to and from the blood

air sac (cut across)

Air sacs

Did you know?

- There are more than 2000 km of blood vessels in your lungs.
- The air sac walls are only one cell thick.

1   Why does your body: a) need oxygen b) have to keep breathing?
2   What is: a) the windpipe b) the bronchus c) an air sac? ▲
3   How do gases travel from one part of your body to another?
4   Why must air sacs have: a) thin walls b) good blood supplies?
5   **Try to find out:** exactly what bronchitis is.

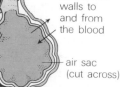

109

8.3 A closer look at the lungs

1 The windpipe is a bendy tube. It has to bend when you move your head. It has rings of stiff **cartilage** (gristle) in it. These rings allow the windpipe to bend, but still keep it in a tube shape. They make sure that it doesn't close up when it bends.

2 The outside surface of the lungs is smooth and moist. This prevents the lungs from rubbing on the chest wall during breathing. The walls of the lungs cannot move themselves, but they are elastic. And so the lungs can stretch and shrink as air flows in and out.

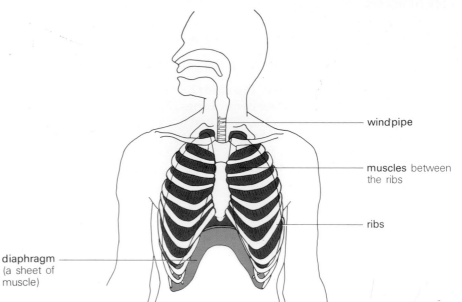

windpipe

muscles between the ribs

ribs

diaphragm (a sheet of muscle)

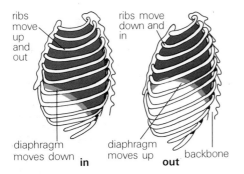

ribs move up and out

ribs move down and in

diaphragm moves down **in**

diaphragm moves up **out**

backbone

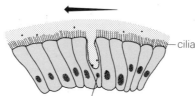

mucus (with trapped dirt and germs is moved along)

cilia

these cells make the mucus

air sac

alveoli

3 Breathing in and out
Two sets of muscles control your breathing. They are the **diaphragm** and the muscles joined to the ribs. When you breathe in, the muscles tighten. This makes the ribs move out and the diaphragm move down. Air flows into the lungs. When you breathe out, your muscles relax. The ribs move in, the diaphragm moves up, and air is pushed out of your lungs.

4 Cleaning up the air
Along the inside of your nose (and windpipe and bronchii) are:

- cells which produce a sticky liquid called **mucus**
- cells with moving hairs called **cilia**.

These cells help to take dirt from the air you breathe in. The mucus traps the dirt. Then the cilia move the mucus along the tubes, away from the lungs.

5 Air sacs and alveoli
There are millions of air sacs in each of your lungs. Each air sac is made up of tiny hollow bubbles called **alveoli**. And so there is a very big surface for gases to get into and out of the blood.
The inside of each air sac is moist. The gas dissolves in the moisture, then passes through the air sac wall.

Did you know?

- The surface area of all the air sacs put together is about the same as the area of a tennis court.
- You either swallow, or cough up, the mucus (and dirt and germs).

1 Why is it important that: a) the windpipe has rings of cartilage in it b) the surface of the lungs is smooth and slimy c) the air sacs have a large surface area? ▲
2 What are 'mucus' and 'cilia'? How do they help to clean up the air you breathe in? ▲
3 Gases can move easily between the lungs and the blood. Give two reasons for this.
4 In what way is an air sac wall like a frog's skin? (see page 16)

8.3 Panting and puffing

How often do you breathe in and out?

There isn't one answer to that question. The rate at which you breathe depends on:

- what you are doing
- how fit you are
- what condition your lungs are in.

You breathe faster when you exercise   When your body is working hard, it needs more energy. It gets this extra energy by 'burning up' more food in the cells. More oxygen is needed to 'burn up' this food. More carbon dioxide is produced, too. And so you breathe faster, changing over the air in your lungs more quickly.

You breathe faster when you are unfit   Normally, you only use a small part of your lungs when you are breathing. Only about ⅙ of the air in your lungs is changed each breath. But when you are exercising, you breathe more deeply.
A well trained athlete may change up to ⅔ of the air in each breath. But if the muscles in your chest are unfit, you will not be able to change over nearly so much air in one breath. This means that you will have to breathe faster to get the extra oxygen you need.

You breathe faster when your lungs are not working properly
Sadly, many people have difficulty breathing because they are suffering from lung diseases like *asthma* and *bronchitis*. Perhaps even more sadly, lots of people have difficulty breathing because they have damaged their lungs by smoking. All sorts of unpleasant things happen to lungs when smoke is breathed into them. The cells which line the lungs' tubes produce extra mucus to try to trap the dirt in the smoke. But the smoke paralyses the cilia, and so all the mucus has to be coughed up. The dust irritates these cells too, and they become swollen. This makes the tubes narrower and breathing more difficult. The air sacs break down, giving a smaller surface for oxygen to get into the blood. It's small wonder that smokers have to breathe fast to get the oxygen they need!

Breathing hard is necessary to get extra energy, quickly!

Superman joins the publicity campaign to show up cigarettes as the harmful things they are

Did you know?

- Around the year 1700, boys at Eton were beaten for not smoking. It was thought that smoking prevented fevers.
- One in three smokers die from diseases caused by smoking.

1 Explain how: a) your body gets extra energy b) your body gets extra oxygen (2 ways)   when you exercise. ▲
2 Why does exercise make you breathe faster when you are unfit? ▲
3 Which diseases affect breathing? ▲
4 What does cigarette smoke do to: a) the cells which line the tubes in your lungs b) the air sacs? ▲
5 Why do smokers cough?
6 **Try to find out:** why 'pollen counts' are made.

All animals need food and oxygen. Even in one day, huge quantities of food and oxygen are used up.

Fortunately, the world's supplies don't run out. Food and oxygen are being made during daylight hours – by green plants! You can't live without them!

Where the food comes from

You eat lots of different foods. But when you trace each food back to see where it comes from, you always arrive at the same kind of answer.

Milk comes from **cows** which live on **grass**.
Eggs are produced by **hens** which live on **grain**.
Bacon comes from **pigs** which are fed on **barley**.

You can trace each food back to a green plant:

Green plants build up the foods upon which animals depend.

comes from / which live on

come from / which live on

comes from / which live on

Energy foods built up by plants

Many of the foods built up by plants are high-energy chemicals called **carbohydrates**. Glucose, sugar and starch are three of these carbohydrates.

Many plants grow stores of carbohydrates. A potato, for example, contains large amounts of starch. You can show this by dropping iodine solution onto a cut-open potato. A blue black colour is produced, the same colour that is produced when starch and iodine are mixed.

These energy stores are important to the plants. The plants use them when they need energy. Plant energy stores are important to you, too. They are used to make many of the energy foods you eat.

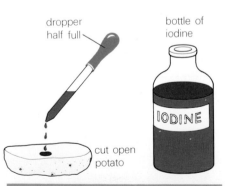

dropper half full

bottle of iodine

IODINE

cut open potato

1 Why does the world not run short of food and oxygen? ▲
2 Why can animals not live without green plants? ▲
3 How would you show that a bean has starch in it?
4 Pick six foods. Trace them all back to see where they come from. Do they all come from green plants?
5 A potato can live all winter in a garden shed. In spring it can grow roots and shoots. Where does it get the energy to do this?

Did you know?

- Peas, beans, wheat and rice contain stores of starch. Grapes contain stores of sugar.
- Almost half the human population's food energy comes from wheat and rice.

You can test a leaf to see if it has food in the form of starch in it by:

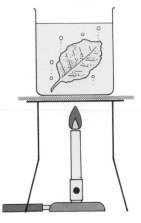

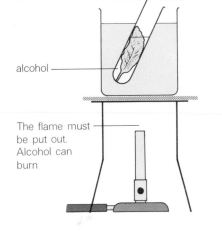

alcohol

The flame must be put out. Alcohol can burn

If the leaf has starch in it, it turns blue-black when iodine is dropped on it

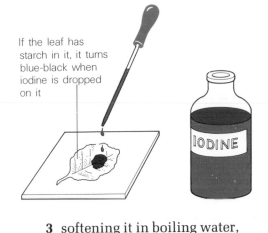

IODINE

1 putting it in boiling water

2 heating it in alcohol (to take away its green colour)

3 softening it in boiling water, then adding iodine to it.

If you use this test on a green leaf which has been growing outside on a sunny day, the leaf goes blue-black. This shows that the leaf *does* have starch in it. But you won't find any starch in:

- a leaf which has been growing in the dark
- a leaf which has been growing in air with the carbon dioxide taken out of it
- a leaf which has had its water supply cut off.

All these facts are true because:

green plants need light, water and carbon dioxide to build up energy foods like starch.

Photosynthesis

Green plants build up energy foods in the process called **photosynthesis**.

photo = using light   *synthesis* = building up

The plants take in carbon dioxide and water and build them into carbohydrates like sugar and starch. Oxygen is produced at the same time. The plants need energy to do this. The energy they use is sunlight. But only plants which contain green chlorophyll can use sunlight. That's why carbohydrates can only be built up by green plants.

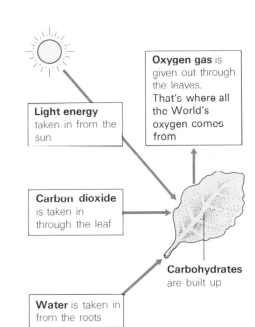

Oxygen gas is given out through the leaves. That's where all the World's oxygen comes from

Light energy taken in from the sun

Carbon dioxide is taken in through the leaf

Carbohydrates are built up

Water is taken in from the roots

1 When you test a leaf to see if it has starch in it, you use alcohol and iodine. What is each substance used for?
2 a) What happens in photosynthesis?
   b) Why does photosynthesis only take place in green plants? ▲
3 Which gas is produced in photosynthesis? ▲
4 Why do aquarium plants often have bubbles on their leaves?
5 **Try to find out:** more about the cells at the surface of a leaf.

Did you know?

- Leaves don't breathe! Gases pass into and out of the cells in the leaf by diffusion.
- Green plants grow better in air which has extra carbon dioxide in it.

8.4   Energy from plants

In a world where new sources of energy are badly needed, scientists are showing lots of interest in green plants. It's not difficult to see why. The energy which green plants supply is useful chemical energy, and it costs nothing to produce. But, most important of all, it can be produced year after year – it is **renewable**.

Scientists have always been interested in plants which produce foods like grain, fruit and vegetables. The whole human population depends on the energy foods which these plants build up year after year. Scientists have helped farmers to produce larger and larger crops from these plants by:

- breeding new plants which produce heavier crops and which resist disease better
- finding out which chemicals help plants to grow well, and making these chemicals into fertilisers
- making chemicals which kill pests.

Fuel producing plants?

But scientists have now begun to investigate fuel producing plants. In particular, they are looking at quick-growing plants which do not need good farming land to grow. You can see four of these plants on this page.

Giant Kelp is a seaweed. It is one of the world's fastest-growing plants. It grows in shallow sea water. Gas and liquid fuels can be made from it

Babussu palm grows in areas of low rainfall. Charcoal, wood gas, liquid fuel and animal food can be made from its nuts

Water hyacinth is a fast-growing weed which grows in fresh water. It can supply gas and liquid fuels

Sugar cane is used in Brazil to make liquid fuel. But sugar cane needs good soil in which to grow. This adds to Brazil's food problems

1 The energy from green plants is 'renewable'. What does this mean? ▲
2 Give the name of one useful chemical which plants build up. What does the plant need to produce it?
3 What is contained in a fertiliser? ▲
4 Why have farmers had to produce greater amounts of food each year?
5 a) The plants in the photographs could all become important for the same reason. What is it?
   b) Which of the plants grow where most other plants normally wouldn't?
6 Can you see anything wrong with: a) using sugar cane to make fuel   b) using chemicals to kill pests?

Did you know?

- Green plants have always supplied fuels. Wood is the main fuel in many poor countries.
- In the U.S.A., alcohol fuel is made from corn.

8.4 Keeping the gases in balance

1 Animals use food to get the energy they need. They use oxygen and produce carbon dioxide, all the time.

2 Plants also use food to get the energy they need. They *also* use oxygen and produce carbon dioxide, all the time.

3 But during the day, green plants build up food! To do this, they use up carbon dioxide, and produce oxygen. Overall, green plants produce more oxygen than they use.

In this way, oxygen and carbon dioxide are constantly being taken from, and added to the atmosphere.

Up till about 200 years ago, the amounts of oxygen and carbon dioxide in the atmosphere stayed roughly the same. The green plants used up all the carbon dioxide produced by the animals. The green plants produced all the oxygen which the animals needed.

But since then, the amounts of oxygen and carbon dioxide in the atmosphere have been changing. There are two main reasons for this.

1 More and more fuels are being burned   In 1980, the amount of fuel burned was four times the amount burned in 1950. Burning huge quantities of fuel uses up lots of oxygen. It produces lots of carbon dioxide too.

2 More and more trees are being cut down   As many as 1/50 of the world's trees may be cut down in a year. And so there are fewer trees to use up the carbon dioxide and to produce oxygen.

For these reasons, the amount of oxygen in the atmosphere is falling. But this isn't a serious worry. The world won't run out of oxygen. The slight increase in carbon dioxide, however, could cause problems. Carbon dioxide prevents heat from escaping from the Earth. If there is extra carbon dioxide in the atmosphere, the world may gradually warm up. This may change the weather. It could even melt the ice at the Poles.

You **use up** the same amount of oxygen as that **made** by a large tree

1 Write down three ways in which oxygen is taken from the atmosphere. ▲
2 Why did the amounts of oxygen and carbon dioxide in the atmosphere stay 'in balance' before 1800? ▲
3 Brazilian farmers clear land for farming by cutting down trees and burning them. How does this change the gases in the atmosphere?
4 Tests show that at night, a plant takes in oxygen and gets rid of carbon dioxide. On a sunny day, it takes in carbon dioxide and gets rid of oxygen. Explain what is happening.
5 **Try to find out:** about the work of the Forestry Commission.

Did you know?

- Trees are the plants which use most carbon dioxide and produce most oxygen.
- The oxygen produced by the grass on a football pitch, is used up by 75 fans in a 90 minute match.
- It needs a forest the size of Greater London to supply enough oxygen to burn the coal or oil used by one power station.

8.5   Food chains

Green plants build up their own food. They are called **producers**.

Animals can't build up their own food. They are called **consumers** (to 'consume' means 'to eat') **Primary consumers** are animals which live only on green plants. **Secondary consumers** are animals which feed on other animals.

A lettuce plant is a producer; a rabbit is a primary consumer; a fox is a secondary consumer. Lettuce makes its own food; rabbits eat lettuce; foxes eat rabbits. This makes up a **food chain**:

lettuce ⟶ rabbit ⟶ fox   (⟶ means 'is eaten by')

secondary consumer

primary consumer

producer

Food webs

But rabbits don't just eat lettuces. And foxes don't just eat rabbits. One food chain can only tell you a little about what different animals eat. **Food webs**, made up of a number of food chains, tell you more:

rose   lettuce   grass   grain

bee   slug   rabbit   chaffinch   doormouse

greenfly

bluetit   thrush

sparrowhawk   fox   owl

1   What is: a) a producer   b) a primary consumer   c) a secondary consumer?   ▲
2   What does a food chain show?
3   About the food web:
   a) Divide the animals and plants into three sets – producers, primary consumers and secondary consumers.
   b) What does a sparrow hawk eat?
   c) A sparrow hawk can help, and make difficulties for, a gardener. Explain why.
   d) Does the dormouse move about mostly by day or night? Explain.
4   **Try to find out:** which foods bees and greenfly get from roses.

Did you know?

- When food is scarce, a fox will eat berries.
- Rabbits occasionally eat worms and snails.

Green plants don't just grow on land. The ocean has green plants, too. The most important of these are the **plant** plankton.

Plant plankton look very different from the plants you see every day. Plant plankton are microscopic. Most are made up of only one cell. But each of them contains chlorophyll. And so each tiny plant does the same job as a green plant on land. It uses sunlight to build up energy foods.

Plant plankton drift about in the sunlit waters near the surface of the sea. Other types of plankton live there, too. These are the slightly bigger **animal** plankton. Many of these animals are tiny shell fish which move about, feeding on the plant plankton.

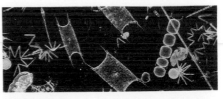
Plant plankton

Animal plankton

Food chains in the sea

The largest amounts of plant plankton are found in shallower parts of the ocean, like the North Sea. That's why most fish are in these parts, too. But fish don't eat plant plankton. The tiny plants are too small for that. What happens is that:

... plant plankton are eaten by animal plankton ...
... animal plankton are eaten by small fish ...
... small fish are eaten by larger fish, which are eaten by larger fish, which are eaten by ...

The sea's food chains can be long. Its food webs can be complicated. But they all start from plant plankton, the ocean's producers.

animal plankton

plant plankton

sand eels

mackerel

cod

herring

haddock

dead organisms fall to the sea bed

crab

shell fish

sea worms

sea bed

A North Sea food web

1   What are plant plankton like? Why are they called producers? ▲
2   Write down some differences between animal and plant plankton. ▲
3   Using the food web, write down: a) some foods eaten by mackerel   b) the fish which eat plankton   c) fish which feed on the bottom of the sea.
4   Why does the North Sea have large stocks of fish? ▲
5   **Try to find out:** what has happened to the herring stocks in the North Sea and why.

Did you know?

- Sometimes parts of the North Sea have so much plankton that the water looks green.
- About 70% of the world's oxygen is produced by plant plankton.

For over 100 years, farmers have regarded the rabbit as a serious pest. It's not difficult to see why. Where rabbits are plentiful, plants suffer badly. The rabbits destroy all sorts of young plants by eating the new shoots. They also eat much grass which could be used for grazing sheep and cattle.

In 1953, large numbers of rabbits were deliberately wiped out by using a disease called myxomatosis. Soon after that, the countryside began to change. As you might expect, the most obvious change was in the plant life. More tree seedlings began to survive. Certain types of wild plant became more plentiful. Grasses which had been cropped short were able to grow to their full height. But the animal life changed, too. Some large plant-eating animals, like the hare and the deer, increased in numbers because there was more vegetation for them to feed on. Other meat-eating animals like the fox, the stoat and buzzard, which fed mainly on rabbits, became fewer in number. They had to look for other sources of food. That's why smaller animals, like mice and voles, began to suffer, too.

All this should help you to understand an important point about food webs:

Anything which affects one part of a food web affects other parts, too.

Sometimes, a change can have rather unexpected effects. You can see this in the examples below.

- In Poland, otters were killed to protect fish stocks. In fact, fish stocks fell.
- DDT, a pesticide, was sprayed on to apple trees to kill pests, including the red spider mite. But it didn't kill the mites. In fact, they multiplied.
- DDT was sprayed on to fields to kill insect pests. Later heron, grebes and other fish-eating birds were poisoned by the chemical.

Why did these things happen?
That's for you to work out in the questions.

Pretty to look at, but a pest to the farmer

When DDT was first used in Britain, the hawk population fell. Now, the use of DDT is strictly controlled

1   Why is the rabbit regarded as a pest? ▲
2   Why did the myxomatosis outbreak affect the numbers
    of: a) buzzards b) deer c) mice? ▲
3   Explain how the unexpected effects mentioned above came about.
    Here are some clues:
        otters often feed off diseased fish which are easy to catch;
        red spider mites are preyed on by other small animals which live in the bark of apple trees;
        DDT stays in the body tissues of any animal which eats it.
4   **Try to find out:** when the rabbit was introduced into Australia, and what effect it had.

Did you know?

- A female rabbit may have six or more litters in one year. She will produce between 3 and 8 young in each litter.
- In 1930, there were thought to be more than 100 000 000 rabbits in Britain.

A revision crossword

Here is a crossword. It is (mostly!) about what you have learned in this section. Copy it into your book and use the clues to complete it.

Clues across

1 Plants build up this food.
6 To get energy, we must ____ food.
7 This gas helps burning.
9 This food comes in a shell.
10 Nitrogen is the most plentiful gas in the air. True or false?
13 This means 'to give out'.
15 You get this noise when you let air out of a tyre.
16 A grain crop.
18 The part of your body where food is digested.
19 Too little energy food makes you ____.
22 Burning magnesium produces ____ energy.
23 The noble gases make up about ____ per cent of the atmosphere.
24 Lime water turns this colour when carbon dioxide is bubbled through it.

Clues down

1 Are lungs rough or smooth?
2 ____ food we eat comes from plants.
3 These food producers lay eggs.
4 The gas produced from boiling water.
5 Plants need this to make starch.
8 A burning splint ____ ____ in carbon dioxide.
11 The ____ cage is round the lungs.
12 Food and air keep ____ alive.
13 The sun's light energy comes from this direction in the morning.
14 You would use this to test for starch.
17 The ____mosphere.
18 Carbon dioxide and water are made when North Sea ____ burns.
20 When we burn up a lot of food we become ____.
21 This is the symbol for a gas used in strip lighting.
22 This is the symbol for a gas used to fill balloons.

Index

Acknowledgements

The publishers wish to thank the following for permission to reproduce photographs:

All-Sport, pp.2, 22 bottom, 23; Animal Photography/S. A. Thompson, p.13 bottom centre; Ardea London/D. B. Burges, p.77 top centre; Ardea London/P. Morris, p.117 bottom; Ardea London/J. Daniels, p.118 bottom; Austin Rover Longbridge, p.34 left; Berry Magicoal Ltd, p.93 centre; The Bettman Archive, p.83 top right; Biophoto Associates, pp.66 right, 67 top, 72; Botany Department, Oxford, p.65 left, right and centre; B.P. Oil Ltd, p.34 top right; British Geological Survey (NERC copyright), p.58; British Rail, p.26 right; Bureau International des Poids et Mesures, p.5 top; Camera Press London/Karsh, p.39 centre; Central Electricity Generating Board, p.93 top; Bruce Coleman Ltd, pp.13 top left, bottom left and right, 70 bottom right, 76, 77 top left, left centre and bottom, 78, 81 top left, right and centre, bottom centre, 114 bottom centre; Controller of HMSO (Meteorological Office), p.43; Derby Art Gallery, p.39 top; Eagle Alexander Communications for the Solid Fuel Advisory Service, p.24 left; Department of Energy Crown copyright, p.22 top centre; C. Feil/Colorific! p. 1 bottom left; Nick Fogden, pp.25, 37 right, 46 top right, 54, 57 top left and right, 60 bottom, 62, 97; Gene Cox, pp. 66 left, 71; B. Harris/Colorific!, p.77 centre right; The Health Education Council, p.111 centre; Hewlett Packard Ltd, p.5 bottom; Eric and David Hosking/D. P. Wilson, pp.114 top right, 117 top; Alan Hutchinson Library Ltd, pp.53, 114 bottom left; London Scientific Fotos, pp.70 top right, 74 top left and right, and bottom right; J. Moss/ Colorific!, p.1 bottom centre; NASSA/Colorific!, p.101 top; Northumbrian Water Authority, p.63; Osram (GEC) Ltd, p.24 top; Oxford Scientific Films: G. Bernard, p.13 top centre; M. Tibbes, p.13 top right; D. Thompson, pp.70 centre, 81 bottom left and right; P. Parks, p.70 left; J. Foott, p.77 top left; p.118 top; Picturepoint London, pp.1 top left and right, 22 top left, 46 top right, 51, 64, 74 bottom left, 86 top, 101 bottom left, 107 top, centre left and right, 114 bottom right, 115; Pilkington Brothers plc, p.38 centre; Redland Prismo Ltd, p.41; Rockware Glass Ltd, p.12; Rolls Royce Ltd, p.1 top centre; Soft (UK) Ltd, p.91 bottom right; Scala/ Vatican Museum, p.83 top left; Seaphot/P. Scoones, p.101 bottom centre; A. Sieveking/ Vision International, pp.73, 75 top and bottom; Barrie Smith/The French Picture Library, p.42; Spectrum Colour Library, pp.27, 32 top right; T. Spencer/Colorific!, p.94; Tony Stone Photolibrary London, p.48; Supersport, pp.24 centre, 111 top; Syndication International, p.22 top right; W & E Vehicles, p.27 bottom right; ZEFA, pp.1 bottom right, 107 bottom.

.... and thanks

We should like to thank

- the readers for considered and constructive criticism of the text
- friends and colleagues for time spent discussing and trying out the material in this book, and for helpful comments and suggestions and, most especially,
- wives and families for long suffering, understanding, support and encouragement, without which this book would not have progressed beyond the first few pages.

AWF
IJG